EINSTEIN'S MOON

Bell's Theorem and the Curious Quest for Quantum Reality

F. DAVID PEAT

CB
CONTEMPORARY BOOKS
CHICAGO

Library of Congress Cataloging-in-Publication Data

Peat, F. David, 1938–
Einstein's moon : Bell's theorem and the curious quest for quantum reality / F. David Peat.
p. cm.
Includes index.
ISBN 0-8092-4512-4 (cloth)
0-8092-3965-5 (paper)
1. Bell's theorem. 2. Quantum theory. I. Title.
QC174.17.B45P43 1990
530.1'2—dc20 90-37370
CIP

Thanks go to John Bell for reading and offering his helpful comments on this manuscript. Any errors or misinterpretations remain, however, the responsibility of the author.

Published by Contemporary Books, Inc.
180 North Michigan Avenue, Chicago, Illinois 60601
Manufactured in the United States of America
International Standard Book Number: 0-8092-4512-4 (cloth)
0-8092-3965-5 (paper)

CONTENTS

What an abyss of uncertainty, whenever the mind feels overtaken by itself; when it, the seeker, is at the same time the dark region through which it must go seeking and where all its equipment will avail it nothing. Seek? More than that: create. It is face to face with something which does not yet exist, to which it alone can give reality and substance, which it alone can bring into the light of day.

—Marcel Proust

We often discuss his notions on objective reality. I recall that during one walk Einstein suddenly stopped, turned to me and asked whether I really believed that the moon exists only when I look at it.

—Abraham Pais

1
THE UNANSWERED QUESTION

What is reality? Do our observations and experiments reveal reality or simply create it? Can the human mind truly understand the universe, or is its deeper meaning forever hidden from us? Questions like these have been asked by scientists and philosophers since the time of the ancient Greeks.

Recently, however, these discussions have taken a remarkable new turn, for physicists are confronting one of the most important scientific discoveries of the last fifty years. This discovery, called Bell's theorem, proposes nothing less than an experimental test of the nature of reality. Already physicists have responded to this challenge and have designed and carried out experiments that suggest the universe is far stranger than anyone ever supposed.

As we reach toward a new millennium, scientists and philosophers are being forced to consider the full implications of Bell's theorem. The result of this new investigation may well be an entirely new form of physics, one that integrates ideas as diverse as quantum theory, general relativity, and the nature of the elementary particles.

Bell's theorem is a work of great imagination, creativity, and ingenuity. It is an elegant way of tricking nature into revealing one of its secrets, a secret that was kept hidden for more than two thousand years. The father of the idea, the Belfast-born physicist John Bell, had long puzzled over the ultimate nature of reality at the level of the atom. His sympathies were with Einstein, who had proposed that an objective reality exists independent of ourselves and our perceptions. Yet, in the 1920s, Niels Bohr, interpreting Werner Heisenberg's new quantum theory, had proposed that such objective reality is a fiction and that our commonsense view of the world does not apply to the world of the atom.

Who was correct, Einstein or Bohr? The debate continued for forty years until, in 1964, Bell discovered a way of forcing nature to reveal the answer. His discovery made it possible to determine, once and for all, whether our commonsense view of the world is correct or profoundly wrong.

The crucial test of Bell's theorem had to wait for the design of some highly sophisticated experiments. The results, however, are now in, and it turns out that the universe is indeed stranger than anyone could have imagined. There is no going back to Einstein's comfortable commonsense world; we are forced to accept a new vision of reality demanded by Bell's theorem. The task now facing physicists and philosophers is to unfold the implications of these new ideas and to create a consistent and coherent account of the world. The story of this search, of the discovery of Bell's theorem and its implications for our conception of reality, is the story of this book.

THE QUANTUM REVOLUTION

Quantum theory represents the most revolutionary change in scientific thinking ever to have taken place. While Einstein's theory of relativity rocked the foundations of physics by proposing entirely new conceptions of space and time—measuring rods that contract close to the speed of light, twins that age at different rates, space-time that can curve in on itself—it was still possible to understand these novel concepts using images and analogies. For example, the idea that time is a fourth dimension may seem bizarre at first, but it is nevertheless within the realm of the imagination; moreover, it can be depicted by drawing a graph. Einstein had stretched common sense to the limit, but he did not destroy it. Even though the universe of space-time was unfamiliar, it could still be conceived of as a part of objective reality. The elements of Einstein's universe were real, and their existence was objective and quite independent of any human observer.

Quantum theory, however, took physics far beyond that limit, for its central mystery lies beyond what can be thought about or said. It is for this reason that quantum theory is so profoundly revolutionary. Its implications were first recognized in 1925, when Heisenberg announced his discovery of a quantum theory. But curiously enough, scientists were not fully confronted with the deepest meaning of quantum theory until, in 1964, John Bell published his proof of what was to become known as Bell's theorem.

The Double-Slit Experiment

To encounter some of the central puzzles of the quantum world, we must take only a small journey. Let us begin with an experiment that leads straight to the heart of the quantum theory and its paradoxes. It is called the double-slit experiment.

The double-slit experiment uses a phenomenon that has been known to physicists for more than two hundred years. It is called the interference of waves. While the idea of wave *interference* may seem unfamiliar, it in fact occurs around us all the time. We see the results of wave interference every time we look at the colored patterns made by oil spilled on the roadway, watch the pattern of ripples on a lake, or enjoy the rich acoustics of a concert hall. Interference takes place whenever waves meet. It doesn't matter what sort of waves they are—sound waves, water waves, light, or even the curious matter waves of the quantum world—they all produce interference patterns when they run into each other.

Imagine that you are standing on a beach looking at ocean waves as they enter the mouth of a harbor. Far out at sea, the crest of each wave stretches out in a straight, undisturbed line. Wave after wave reaches the harbor, enters through its mouth, and spreads out again.

FIGURE 1-1

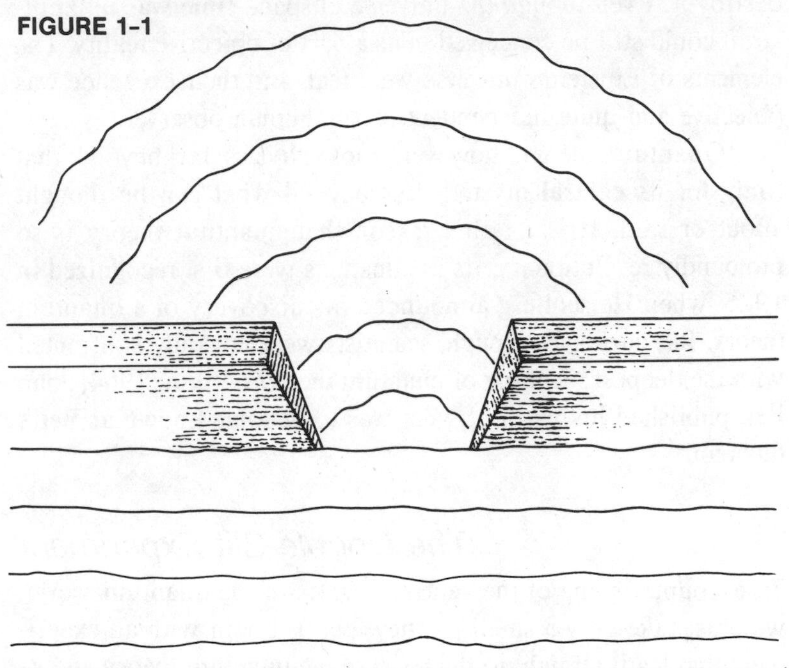

Ocean waves enter the narrow gap in a harbor mouth and spread outward.

Imagine, now, that the harbor has two such gaps or mouths occurring side by side. The waves spread out from the two gaps and meet to interfere with each other. When two crests happen to encounter each other, their effects add up to produce a wave of even greater height. The meeting of two troughs results in a deeper depression. What happens when a trough and crest happen to coincide? The effects of the waves are canceled out, producing a tiny region of relative calmness. As the spreading waves from the two gaps meet and interfere, a complicated pattern of peaks, troughs, and calm water is created.

FIGURE 1-2

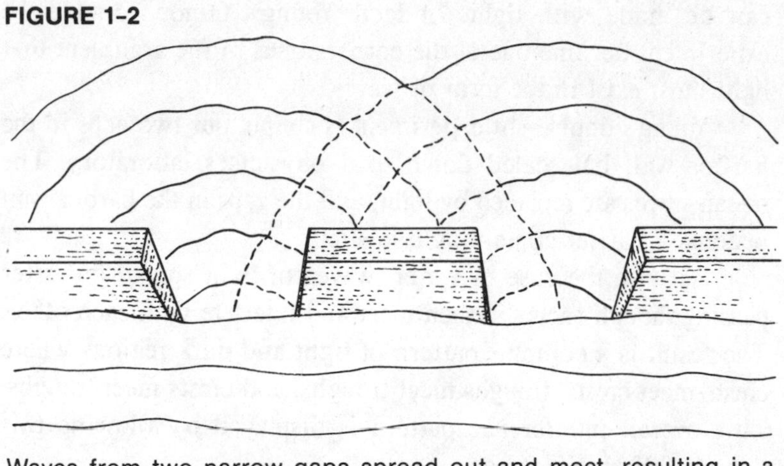

Waves from two narrow gaps spread out and meet, resulting in a complex interference pattern of many peaks and troughs.

Something similar happens in a concert hall when sound waves from a piano spread out and bounce off the walls and ceiling until, by a variety of different paths, they reach your ear. In the region of your head, many reflected waves meet and interfere to produce the rich pattern of sound we associate with the acoustics of that concert hall.

The interference of water waves had long been known, but it was the English physicist and doctor Thomas Young who, in 1801, showed that exactly the same sorts of interference patterns

FIGURE 1-3

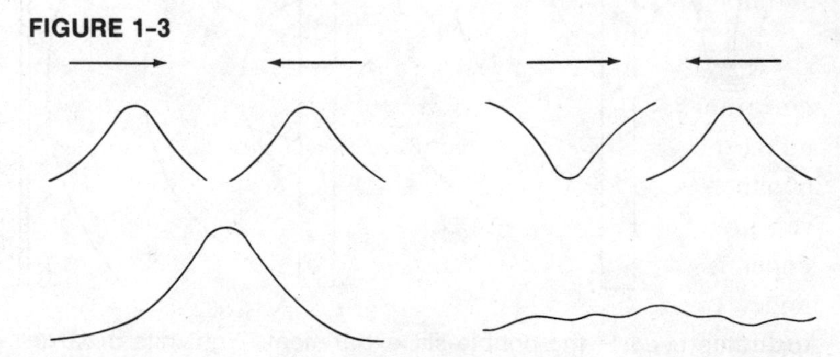

On the left-hand side, two peaks meet and produce an even higher peak. On the right, a trough meets a peak and produces a region of flat water.

can be made with light. In fact, Young's famous double-slit experiment became one of the cornerstones in the argument that light must exist in the form of waves.

Young's double-slit experiment is simply our two gaps in the harbor wall but scaled down to a physicist's laboratory. The ocean waves are replaced by light, and the gaps in the harbor wall become a barrier containing two slits.

As with the case of water, waves of light spread out after passing though each slit, then meet and interfere with each other. The result is a complex pattern of light and dark regions where crests meet crests, troughs meet troughs, and crests meet troughs. This overall interference pattern is displayed by allowing this light to fall onto a screen.

FIGURE 1-4

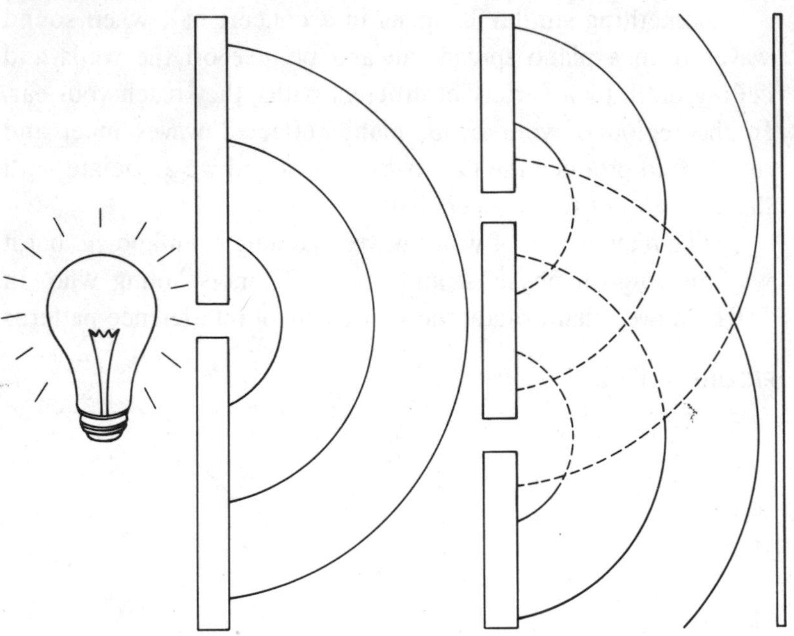

Apparatus used in the double-slit experiment. Light hits a barrier containing two narrow slits. The light passes through the double slits and spreads out to produce an interference pattern, which is registered on the screen.

FIGURE 1-5

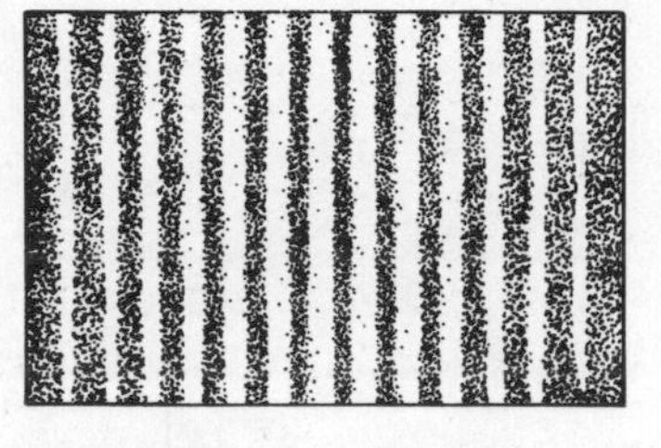

Interference from the double slits produces an alternating pattern of light and dark bands on the screen.

It is easy to see that this interference pattern can be produced only by waves, not by particles. Suppose that in place of a beam of light we have a gun that fires tiny bullets toward the double slits. While most of the bullets will hit the barrier, some lucky ones will pass through one or the other of the slits and carry on until they hit the screen. As far as any one bullet is concerned, it doesn't much matter where the others go or which slits they pass through. The path of each bullet is independent of the paths of all the other bullets, so the result is a simple pattern of tiny holes on the screen—something very different from the interference patterns of light. Of course, since no gun has absolute precision, the pattern of hits will be slightly spread out, as the illustration shows, but nevertheless very different from a wave-interference pattern.

If light consisted of tiny corpuscles, as the great Isaac Newton believed, then, like bullets, they would form a bunched-up pattern directly behind each slit. By contrast, since light consists of waves, its interference pattern spreads out across the whole screen so that light is found even in those regions of the screen that do not lie directly behind the slits. The reason is that light waves can spread out from each slit and interfere with each other. In other words, the interference pattern is the result of crests and troughs coming from all possible different directions.

This interpretation of wave interference can be confirmed by blocking one of the two slits. Light waves will still spread out from the single unblocked slit, but they will no longer encounter a second set of waves. Since interference does not take place, the

FIGURE 1-6

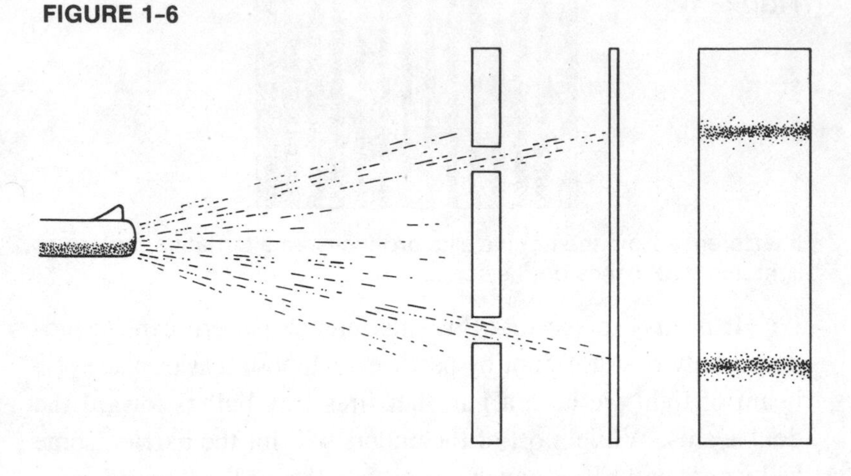

If tiny corpuscles are fired at a double slit, they leave a simple pattern on the screen.

characteristic light and dark pattern on the screen vanishes; in fact, the resulting pattern is closer to that produced by corpuscles or tiny bullets. Clearly, interference must be the result of waves that pass through both slits. At least this is what everyone believed until quantum theory came along!

The Quantum Double Slit

Armed with the classical or commonsense double-slit experiment of Thomas Young, it is possible to enter the quantum world. We begin by gradually reducing the intensity of the beam of light that hits the double slits. The immediate result is that the interference pattern on the screen gets weaker and weaker until it can no longer be seen with the naked eye. But this is no problem, because we can attach a piece of photographic film to the screen and leave it there for several hours, or even for days, so that the very weak pattern has time to build up. It turns out that no matter how weak the beam of light when the film is developed, we always find the same interference pattern.

But what happens if the intensity of light is cut down to such a point that we are forced to take into account its quantum

FIGURE 1-7

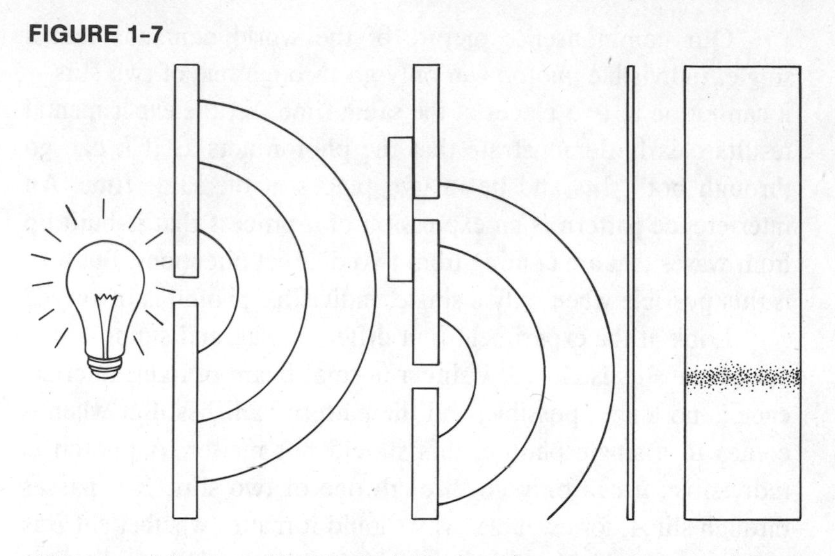

With one of the two slits blocked, the interference pattern vanishes, and a single strip of light is seen on the screen.

nature? Quantum theory teaches that light is ultimately made out of finite and indivisible quanta called photons. What happens when the light is so weak that just one photon is reaching the double slits at any time? Common sense dictates that, since a photon is indivisible, it can go through only one of the two slits. It must therefore pass through either slit *A* or slit *B* and then hit the screen. Since the photon goes through only one slit, an interference pattern is clearly impossible! In fact, when single photons are involved, the pattern should look the same as that produced by firing tiny bullets. Common sense dictates that the interference pattern should vanish. ***It is exactly at this point that the quantum world parts company with logic and common sense.***

Using very delicate apparatus, scientists have been able to carry out the double-slit experiment using only one photon at a time. When the results of a very large number of these individual events are collected, the familiar interference pattern appears. It is exactly as if waves of light have been passing through both slits, meeting up, and interfering. Yet only a single photon was present at any one time!

Our commonsense picture of the world demands that a single, indivisible photon can only go through one of two slits—it cannot be in two places at the same time. Yet the experimental results clearly demonstrate that the photon acts as if it can go through both slits and be in two places at the same time. An interference pattern is an expression of a process that is built up from waves that are coming from two distinct directions. But how is this possible when only a single, indivisible photon is involved?

Look at the experiment in a different way, and suppose that one of the slits is closed. Using a normal beam of light, interference is no longer possible, and the pattern vanishes. But when it comes to a single photon, this should not matter. A photon is indivisible; it can only go through one of two slits. If it passes through slit *A*, for example, why should it matter whether slit *B* is open or closed? Nevertheless, it does matter. When slit *B* is closed, the quantum-interference pattern vanishes; when slit *B* is open, this pattern reappears. But how can this be? How can a photon passing through slit *A* know whether slit *B* is open or closed? In passing through one part of the screen, how can it know what is happening in some other part?

Could the photon somehow split itself in two, so that half a photon goes through slit *A* and half through slit *B*? No, photons are by their very nature indivisible, so they cannot split in half. The inevitable conclusion is that quantum theory has thrown common sense out of the window and that physicists have been forced to acknowledge that single, indivisible photons *act* as if they can pass through two slits at once and be in two places at the same time. Either that, or some new and mysterious communication is taking place that seems to inform a photon in one part of the universe what is happening in other parts.

PARTICLES OR WAVES?

At the quantum level, light seems to behave as if it could be in two places at once, or as if it were party to some bizarre form of communication. Already, at this early stage of the argument, quantum theory is forcing us to accept the possibility that there could be a nonlocal element to the universe, that quantum particles may not be confined by the everyday, commonsense concept of locality. We are used to thinking of objects as being localized in space, but suppose that this is no longer true. Suppose objects also have a nonlocal aspect—an ability to be in touch with events that are distant from them.

If all were not sufficiently bizarre, it turns out that exactly the same thing takes place with quantum particles of matter called electrons. Physicists have repeated the double-slit experiment using beams of elementary particles whose intensity can be cut down until individual events are registered. A wavelike interference pattern persists. Yet particles are localizable; they have definite edges and are totally different from waves. How can matter behave in such a schizophrenic way? Part of the answer was given by Louis de Broglie.

During the 1920s, de Broglie, a young Frenchman, unable to decide between the traditional family career of diplomacy and his younger brother's pursuit of physics, entered the Sorbonne to carry out research on both topics. While writing his thesis on French history, he was struck by the revolutionary idea that matter may possess a dual nature—that it could be both a wave and a particle. His doctoral thesis, written in 1924, suggests that the electron, in addition to being a localized particle, could also exist as a "matter wave."

Initially, de Broglie's ideas were not taken seriously by physicists, although Einstein himself did try to extend the Frenchman's hypothesis. However, in 1927 two U.S. physicists, Clinton Davisson and Lester Germer, followed by the Englishman George Thomson, decided to test de Broglie's curious idea by repeating the double-slit experiment, this time using a beam of electrons in place of a beam of light.

To carry out the experiment, it was necessary to create a scaled-down version of Thomas Young's original apparatus. The details of their apparatus don't really matter here; the important result is that the experimentalists observed exactly the same sorts of interference patterns that characterized light and other waves.* But this meant that each electron, each quantum particle of matter, must be spreading out from the slits, meeting other electron-waves, and experiencing interference.

Until de Broglie came along, everyone was perfectly happy to accept the electron as a particle, as an incredibly tiny ball that occupied a definite place in space. Scientists knew that electrons could be shot at each other and that they bounce and scatter one off the other like billiard balls. But how could a tiny ball, localized in space, act like a spreading and interfering wave? While many other experiments had demonstrated the particle nature of the electron, Davisson and Germer's results pointed irrefutably toward waves. The only conceivable conclusion was that the electron has a dual, schizophrenic existence: Design an experiment that looks for particlelike behavior, and the electron behaves like a particle. Design one that looks for waves, and it behaves like a wave!

In fact, all quantum particles appear to behave in exactly the same dual way. Electrons, protons, neutrons, the whole quantum "zoo" of elementary "particles" will at some times behave like waves and at other times like particles. Even light itself can behave like a particle, for it is perfectly possible to bounce a photon—a localized particle of light—off an electron, and in this case it acts just like a particle! The quantum world has a twofold nature in which everything is both a particle and a wave.

The quantum double-slit experiment illustrates the first of the great mysteries of the quantum world: the schizophrenic dual nature of matter and energy, in which light and electrons can behave both like particles and like waves. It also brings us face to

---

*The wavelength corresponding to the matter wave of the electron is enormously short—so short in fact that its interference effects could never be detected using conventional slits and a screen. The two physicists were therefore forced to use an ingenious trick. Atoms in a metal arrange themselves in a regular pattern, or lattice. Davisson and Germer made use of the gaps between each of these lattice planes, which act in the same way as the slits in Young's traditional experiment.

face with the essential paradox that a single, indivisible quantum object behaves exactly as if it has passed simultaneously through two separate slits—it can be in two separate places in the universe!

The Mystery of the Quantum Jump

Quantum theory makes yet another radical break with our commonsense approach to reality: not only does a quantum particle seem to be in two different places at the same time—it can also move between two points without ever occupying the intermediate space between! This can be seen in the phenomenon of radioactivity and what is known as the quantum jump.

Our life on earth—indeed, the very existence of the universe itself—is dependent upon the stability of the nuclei (the centers) of the atoms that make up our bodies and all the things around us. In fact, nearly all the chemical elements are absolutely stable, but a few naturally occurring radioactive elements do exist. Radium, that remarkable discovery of the Curies, happens to have one of these unstable, radioactive nuclei.

Imagine a sample of radium—a collection of unstable nuclei—surrounded by a set of Geiger counters. For a long time nothing happens, then suddenly an individual nucleus disintegrates, firing off a particle in the process. But what exactly causes the nucleus to shoot out an elementary particle?

In a radioactive substance like radium, the nuclei have excess energy and are therefore energetically unstable—they are like time bombs just waiting to go off. An elementary particle sitting inside a radioactive nucleus has too much energy for its own good. It wants to escape, to shoot outside and take that excess energy with it. Put another way, the nucleus of radium 228 is like a mousetrap waiting to snap. The problem is that there is no atomic mouse around to trigger the radium mousetrap.

The radioactive nucleus is a little like a ball resting in a tiny valley high on top of a mountain. The ball is unstable; it has too much energy, and if it is pushed to the top of its valley, it will roll down the mountain at greater and greater speed and in this way

FIGURE 1-8

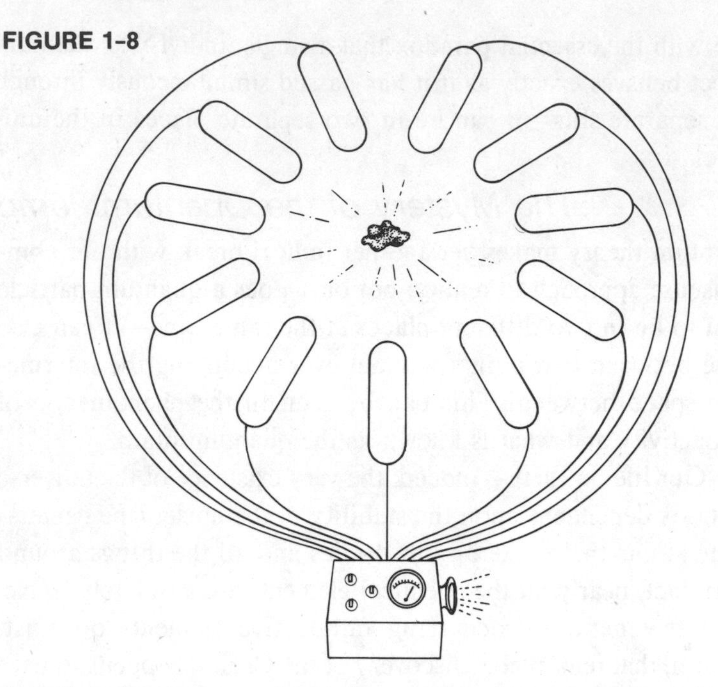

A radioactive isotope sits in the middle of a collection of Geiger counters. If the nucleus disintegrates, the event will be registered on one of the counters.

get rid of its excess energy. Yet, in another sense, the ball is stable, for it requires that initial push to get it to the top of its valley. Indeed, barring human intervention or a very high wind, the ball will remain in its energetically unstable position forever.

A ball sitting in such a valley will never spontaneously find itself rolling down the mountainside. If it is left to itself, such a sudden change is impossible—a solid wall of rock and earth separates the ball in the valley from the side of the mountain. Likewise, for an elementary particle in a radioactive nucleus, common sense dictates that an isolated nucleus can never decay. The elementary particle requires a tiny kick of energy in order to get it to the lip of its energy valley. From that point on, it can shoot out of the nucleus, carrying its excess energy.

In our everyday world, things happen for a reason and according to a well-defined sequence of events. A pet mouse escapes from its cage by wriggling through the bars. This is a perfectly comprehensible escape plan. There is a definite time when the mouse is in the cage, a definite time after it has escaped, and a time in between when the mouse is in the process of wriggling between the bars.

FIGURE 1-9

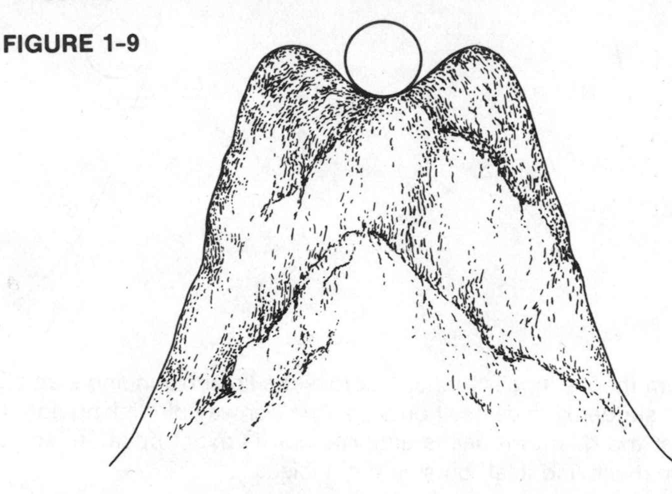

A ball balanced on a knife-edge is absolutely unstable; a puff of wind will topple it. This ball also is unstable, but it requires a push to move it out of its depression so that it can roll downward.

At this point, quantum theory parts company with everyday analogies. Quantum theory dictates that radioactive atoms cannot be understood in this commonsense way. In place of a continuous change is a discontinuous leap. At one instant, the elementary particle is inside the nucleus. At the next, it has escaped. There is no intermediate state, no time in which the particle is actually in the process of getting out. Unlike a mouse, a quantum particle will never be discovered with its head poking out and its tail sticking in!

Quantum theorists call this discontinuous transition the quantum jump. An instant before the jump, the elementary particle is occupying a given region of space. An instant later, it is somewhere else. And, according to the quantum theory, no physical process connects these two states of being, no duration of

FIGURE 1-10

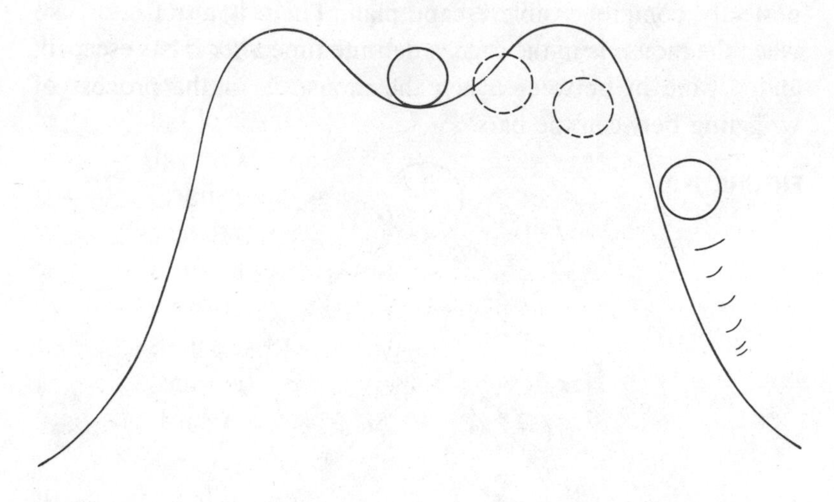

In quantum theory a ball "tunnels" across the lip surrounding a small valley and suddenly finds itself outside. Another way of thinking about this is that the quantum ball is uncertain of its exact position, so at some time it will find itself outside the nucleus.

time separates them.* It is as if the elementary particle suddenly flickered out of existence, passed through a limbo of "no time" and "no space," and then reappeared somewhere else.

This is a staggering reversal of "commonsense" reality. I throw a ball toward you. It leaves my hand, and a moment later you catch it. In the short space of time between the ball leaving my hand and it being caught by yours, we both know that the ball must be *somewhere*; it must be traveling through the air, occupying space, and moving in a trajectory through space and time. Not so in the quantum world. At one instant the particle is inside the nucleus; in the next it is traveling away at high speed. Nothing happens in between. This is the mystery of the quantum jump. What sort of a "reality" can accommodate such a notion?

*Textbooks on quantum theory talk about such things as "quantum tunneling" and the leakage of the wave function out of its energy valley. But in the last analysis, these are not really explanations of what happens, for they give no account of the nature of the quantum jump itself.

Indeterminacy

The radioactive nucleus holds yet another mystery, for it turns out that the quantum jump involved in its disintegration is totally unpredictable. It is possible to sit and watch, through a microscope, the disintegration of nuclei in a tiny piece of radium. Here and there across the sample, tiny bursts of lights can be seen. Each bright green flash of light represents the disintegration of a nucleus. But the whole pattern occurs at random, and there is no way of predicting where or when the next flash will occur. The instant of a quantum jump is an absolutely random event.

At first sight, there may seem to be nothing unusual about this. After all, chance events are with us all the time. Tossing a coin, playing roulette, betting on the draw of a card, and even being hit by a car are all chance events.

However, we do know that all of these processes are, at heart, deterministic. When a coin lands heads, we know that something must have caused it to land in this particular way. The motion of a coin through the air is a complicated business; nevertheless, every millimeter of its path is totally determined, albeit by a host of complex causes such as tiny air currents. Finally the coin hits the table, spins, and bounces. This complex motion also depends on many things—on the exact angle at which the coin hits the table, its speed, the rapidity of spin, the friction of the tabletop, the roughness of the coin, etc. But all these different causes add up to the final state of the coin.

In practice, it would never be possible to pin down and measure, with total accuracy, this myriad of tiny effects and thereby arrive at an exact prediction of the result of an individual coin toss. But at least it is reasonable to assume that, over a sequence of coin tosses, these effects will tend to average themselves out, so that there will be a 50 percent chance of getting heads and a 50 percent chance of getting tails.

We may deal in probabilities when it comes to roulette wheels and coin tosses, but the process is still deterministic. Suppose that godlike beings existed who had perfect knowledge of all the air currents in a room, the exact roughness of each part of the table, the precise speed at which the coin toss was made, etc. If all the multitudes of variables acting on the spinning coin could be known exactly, then indeed it would be possible to predict the outcome of a single coin toss using a high-speed computer. Nothing within the laws of classical physics prevents us from predicting the outcome of deterministic events; it is simply their complexity that causes us to resort to probabilities.

Probability, in our everyday world, is a mathematical way of dealing with our ignorance about the details of complex causes by assuming that things tend to average out in the end. The weather is a good example of such a probabilistic system, since a great many complex effects act together to create tomorrow's weather. Fifty years ago, our ignorance about the physics of weather, as well as our inability to collect accurate data, was so great that weather forecasts were given in a fairly informal way: "It will be sunny tomorrow with a chance of rain"—little could be said about the days that followed. But now we are told that the high temperature will be 68° tomorrow with a 20 percent chance of rain, the chance of rain increasing to 40 percent by the weekend. The use of weather satellites, increasingly accurate monitoring, and high-speed computers to work out weather models has made it possible to attack our areas of ignorance and in this way give much more accurate predictions, in terms of more refined probabilities. By gathering ever more detailed data and by studying the nature of a variety of causes, it becomes possible to predict individual events in our large-scale world with greater and greater accuracy.

The disintegration of a radioactive nucleus is profoundly different. Quantum theory dictates that the probabilistic answers are absolute and irreducible. They are not a measure of ignorance but of absolute chance. It is not a matter of being ignorant of exact details, as in the coin toss, for there are no hidden causes working at the quantum level; no missing data are waiting to be discovered. To put it bluntly, there *are* no processes that make the nucleus disintegrate. As we have already learned, a quantum jump simply happens. No one can predict the exact moment when a radioactive nucleus will flash with light. All physics can do is provide a set of betting odds—for example, a 50 percent probability that the event will happen within the next hour.

Take, for example, a sample of radium 228, which contains a large number of radioactive atoms. Quantum physics dictates that we can never know exactly when one of these nuclei will disintegrate. Nevertheless, we do know that in 6.7 days, half these nuclei will have decayed and half will not. In a further 6.7 days, the half remaining will have decayed, and so on. We know the *average* half-life of radium nuclei. We know that at any moment some nuclei will be disintegrating while others remain stable—yet all we can do is to give probabilities, for quantum theory asserts that nothing we can ever do will improve upon these odds. Quantum probabilities are absolute, not a measure of ignorance. Quantum events, the theory teaches, are indeterministic in origin.

Yet how, we wonder, does a nucleus know that "its time has come"? What causes one nucleus to decay while another remains stable? Common sense demands an answer to these questions. But is this yet another area of the quantum world in which common sense has to be abandoned?

For two hundred years, physicists have been searching for processes, mechanisms, and causes within the world around them. Now quantum theory is saying that, at the level of quantum processes, no such hidden causes exist; the breaking apart of a radioactive nucleus will never be predictable, and quantum physics must live with betting odds and probabilities.

This bleak conclusion was never accepted by Einstein, who countered, "God does not play dice with the universe." The great physicist was not willing to accept that nature's most fundamental processes occur by pure chance. Nevertheless, quantum theory and the implications of decades of quantum experiments firmly reject Einstein's belief. Nature at its most fundamental level is indeterministic and probabilistic.

CONCLUSION

We have come face to face with the great quantum mysteries:

- The duality of light and matter, in which quantum objects can be both waves and particles, local and nonlocal
- The double-slit experiment, in which a single, indivisible quantum particle acts as if it has been in two different places at the same time
- The quantum jump, in which change takes place in a sudden and discontinuous fashion with no state between
- The absolute and irreducible indeterminism of quantum processes—God playing at dice

2
QUANTUM THEORY

FROM RUTHERFORD TO BOHR

The first years of this century were truly revolutionary for physics. In Switzerland, Albert Einstein announced his special theory of relativity. In Paris, Pierre and Marie Curie discovered two new elements, polonium and radium, and were working on the curious phenomenon of radioactivity. In Manchester, England, Ernest Rutherford was trying to figure out the nuts and bolts that hold the atom together.

By 1907 Niels Bohr, a twenty-two-year-old student who was to become one of the key players in the quantum world, had already received a gold medal from the Royal Danish Academy of Sciences and Letters for his first piece of research. The young Bohr was keeping in touch with the latest developments in physics, and soon decided to leave his native Copenhagen and travel to Manchester to find out, firsthand, what Rutherford was working on.

Ernest Rutherford was one of the truly outstanding physicists of all time, a man who in his abilities has been compared to the great experimental physicist Michael Faraday. Born in New Zealand, he was bursting with energy. Sometimes boastful, always larger than life, Rutherford was an exceptional experimentalist with a deep intuition about the inner workings of nature. His laboratory at Manchester was a dynamo of activity. It is said that when things were going well, his students would hear the great man singing "Onward Christian Soldiers," but if an experiment did not work out, then "Fight the Good Fight" would come booming down the corridors of the physics department.

Rutherford already knew that the atom contains smaller parts—tiny, negatively charged particles called electrons and much (two thousand times) heavier positively charged particles called protons. Rutherford also knew that an equal number of electrons and protons make up the atoms around us. (Some time later, physicists would also discover neutrons, mesons, and yet other particles forming the atom.)

The question Rutherford had posed himself was: how are these tiny charged particles arranged within the atom? Are they just dotted about like fruit in a cake, or does the atom have some more regular structure?

FIGURE 2-1

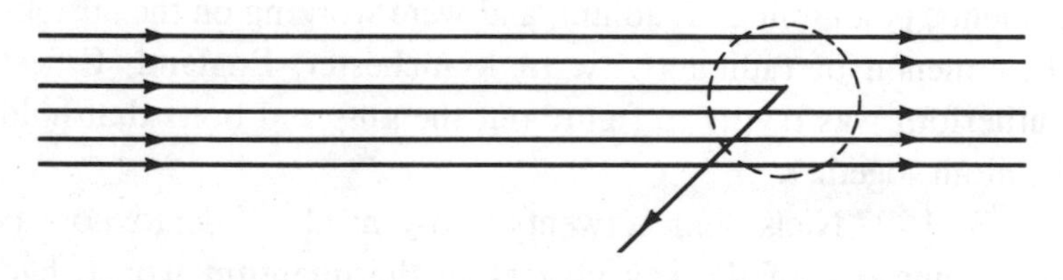

Rutherford fired tiny particles into the center of an atom. He discovered that most of them passed through undetected. Only a very few appeared to hit a hard center and recoil back. Rutherford concluded that most of the atom is empty but that it has a relatively massive central core (nucleus).

The problem was that the atom is far too small even to be seen through a microscope. Rutherford puzzled over the problem and hit on the ingenious idea of firing atomic particles themselves into the atom. Each time one of his atomic bullets hit a charged particle in the atom, it would be deflected. By correlating the pattern of deflections in many experiments, he would be able to build up a picture of how the electrons and protons were packed together inside of an atom. It was a little like collecting litter after a college football match and working out where most of the crowd had been standing.

The data Rutherford collected proved to be a considerable surprise. Most of his atomic bullets flew through the atom undeflected, with only a few of them bouncing back. The reason was that, instead of electrons and protons floating about inside the atom like fish in a bowl, all the heavy protons were concentrated within a very tiny center called the nucleus. In fact, atoms were mostly nothing at all—just empty space. An atom has a very dense central nucleus of protons, in which almost all its mass is concentrated, and around which floats a very thin cloud of electrons.

It turned out that each time Rutherford fired one of his atomic bullets, it had very little chance of hitting anything at all. With every ten thousand bullets fired, only one or two were lucky enough to hit the nucleus and bounce straight back. The rest simply missed the bull's-eye and shot through the atom, undeflected. By 1911 Rutherford was able to figure out that almost all the mass of the atom was concentrated in a central nucleus that was around ten thousand times smaller than the atom itself!

It was at this moment that Niels Bohr arrived on the scene. Whereas Rutherford had been able to build up a picture of the atom experimentally, would Bohr now be able to make theoretical sense of this picture?

The task that faced the Danish physicist turned out to be as important as the one that had faced Newton two hundred fifty years before, when he attempted to explain the motion of the planets. Just as Bohr was able to draw upon Rutherford's new picture of the atom, so Newton before him had used Kepler's picture of the way planets move around the sun and Galileo's experiments on pendulums and falling bodies. What Newton had to do was to take all these different pictures and concepts and tie them together within a single, unified theory of the universe. Similarly, Bohr sought to integrate all that was known about

FIGURE 2-2

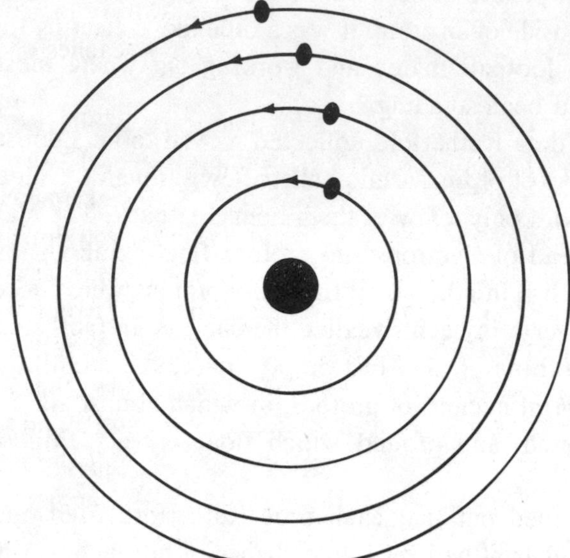

Rutherford pictured the atom as containing a massive central core around which electrons orbited.

physics in the first years of the twentieth century and piece it together into a theory of the atom.

What he could not realize was that, just over a decade later, the picture of the atom that would emerge would require a profound break with our commonsense idea of reality.

Rutherford's conception of electrons moving around a central nucleus looks very similar to a scaled-down version of our solar system. At the center of our solar system is the sun, and around it rotate the planets, each in its own orbit. As with Rutherford's model of the atom, nearly all the mass of the solar system is concentrated in a central nucleus—the sun. The major difference between atoms and planets appeared, at that time, to be one of size. In addition, an electrical attraction, not the force of gravity, keeps the planetary electrons in space.

This picture of a "solar-system atom" suggests that nature repeats itself on different scales and that the same laws of physics apply at the atomic and the astronomical scale. The idea was a particularly beautiful one, for it implied a unity of nature that extends from below the atom to the farthest reaches of the universe.

This model of electrons in planetary orbits would also explain another puzzle about the atomic world—the various characteristic atomic colors, or spectra, that experimental physicists were busy working on. Bohr knew that if you throw a pinch of common salt (sodium chloride) into a lit candle, it produces a bright yellow flame. Stare into a campfire, and you notice pale lilac flames produced by potassium in the wood. When burned, each element produces its own characteristic color. Scientists had already analyzed these colors in greater detail and found that each is made up of a whole series of narrow lines called a spectrum. Each line in the spectrum is associated with a given frequency (or energy) of light. The spectrum of each of the elements is like a fingerprint, with no two different elements having the same spectrum or pattern of lines. The spectrum of a particular atom can also be compared with the series of notes in a musical scale, with each of the elements having its own unique scale.

FIGURE 2-3

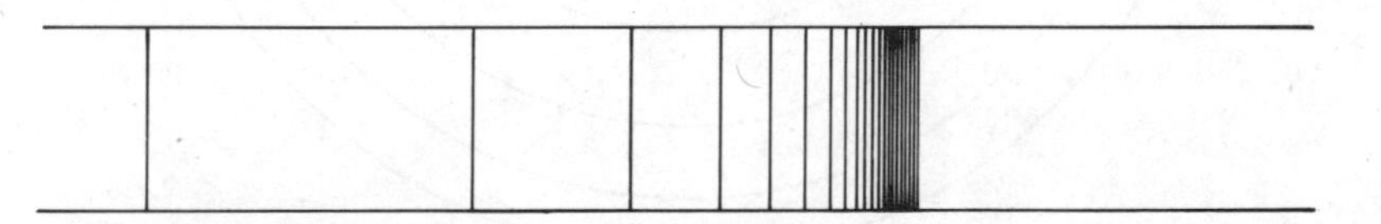

This is an illustration of the atomic spectrum for hydrogen.

Bohr realized that Rutherford's solar-system model of the atom would be able to explain the spectra of different atoms. Each atom has its own pattern of "planetary" orbits. And just as each of the planets in the solar system has its characteristic energy, so the energies of the electronic orbits would be unique. Bohr argued that when an electron moves from one orbit to another, it will give out a burst of energy in the form of light of a particular color. By jumping through a series of orbits, the atom will emit a whole series of lines—in other words, a spectrum. Spectra are the direct result of planetary electrons hopping from orbit to orbit around the nucleus and emitting or absorbing light-energy. Since all the atoms of one element have a characteristic set of orbits, different from the orbits of some other element, the corresponding spectra provide a unique "fingerprint."

FIGURE 2–4

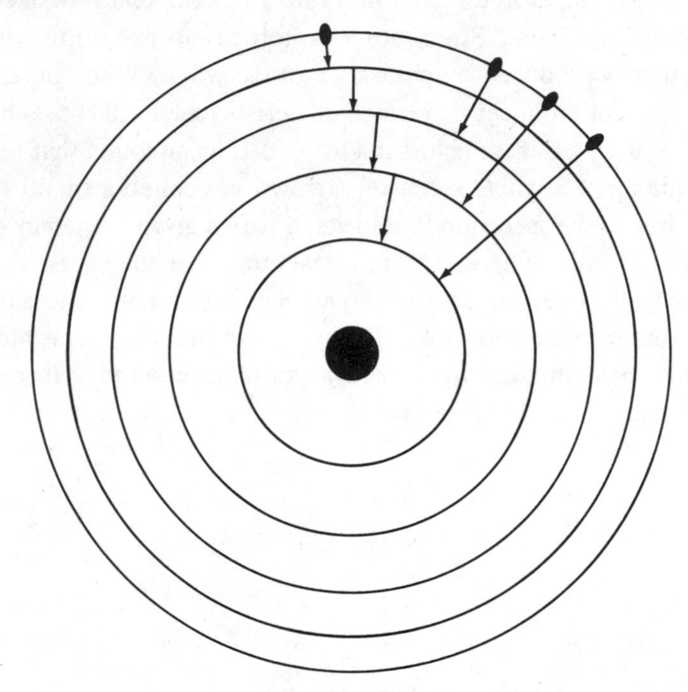

An atomic spectrum (see illustration on page 25) is produced as an electron falls inward, from orbit to orbit. With each quantum jump, the electron emits a quantum of light. Since the outer orbits are more closely spaced together, a spectrum is produced in which lines bunch up at one end. The particular spectrum is a fingerprint of all atoms belonging to a given element.

So the answer to the secret of the atom seemed straightforward. It was simply a miniaturized solar system obeying a set of laws that extended from the stars to the electrons and protons. There was really nothing strange about atomic reality.

However, Bohr knew that, on the basis of classical physics, there had to be something profoundly wrong with such a picture. It would be nice to believe that nature is unified, with the same structures appearing at each scale. But the laws of physics that were known in the early twentieth century simply would not allow the atom to exist in this way. Bohr knew that the problem with this solar-system atom was that planetary electrons are not held in orbit by gravitational attractions but by electrical forces, electrons being negatively charged and protons having positive charges. Physicists already knew what happens when charged particles travel in orbits: they radiate away energy. In a similar way, a charged electron rotating around a positive nucleus would act as a miniature transmitter, radiating away its energy. With each trip around the atom, each electronic "year," some of the electron's energy would leak away.

It takes only a blink of an eye for the electron to make hundreds of millions of orbits around its nuclear sun. Long before that time had expired, the electron would have lost all its energy and spiraled inward to hit the nucleus. Bohr realized that, according to the laws of physics, atoms therefore should not exist even for the blink of an eye—and, since everything is made out of atoms, the entire universe should have disappeared within a split second of its creation.

Yet matter definitely does exist. Something strange must be at work, for, in apparent defiance of the laws of nature, atoms are stable and have persisted longer than the earth itself. How is it possible, Bohr asked, to explain this extraordinary stability of atoms? How have electrons and protons conspired to break the laws of physics and prevent energy leaking away?

The answer, Bohr realized, had to lie in something totally new. There had to be an additional clue that would unlock the mystery of the atom. Bohr had already brought together Rutherford's solar-system picture of the atom with the atomic spectra that physicists were currently analyzing. But there had to be a third component, some missing piece of the atomic jigsaw puzzle. The answer, Bohr realized, lay in that bizarre new idea that Max

Planck and Albert Einstein had been puzzling over.

A few years earlier, Planck and Einstein had been investigating the ultimate nature of light. Light had always been something of a puzzle in physics. In the eighteenth century, the great Isaac Newton had argued that because light travels in straight lines, it must be made out of tiny "corpuscles." His contemporary, the Dutch physicist Christian Huygens, argued, however, that light exists in the form of waves. Thanks to Thomas Young's later experiments, physicists had come to accept this latter picture. But now, in the first years of the twentieth century, Planck and Einstein were claiming that, at its ultimate level, light exists in the form of tiny, discrete packages called quanta. These quanta are indivisible, so there is no such thing as half or any other fraction of a quantum.

Because these quanta are so small, we, in our large-scale world, see light as continuous waves. But in the very smallest scales, light has a grainy nature. In fact, it was beginning to look as if all energy takes a quantized form. Any process in which energy is taken in or given out takes place in a series of "quantum jumps" as quanta of energy are exchanged.

This, Bohr realized, was the key to Rutherford's atoms. The clear implication of the quantum concept is that energy cannot gradually leak away; it can only be given out in finite packages. The energy of an electron as it orbits around the nucleus cannot leak or radiate away; it can only be given out in a finite amount. Electrons cannot gradually spiral inward, Bohr reasoned; they can only jump from orbit to orbit. An electron-planet changes its orbit only by losing or collecting a whole quantum of energy.

The atomic picture suddenly came into focus. An electron, Bohr argued, cannot lose energy by spiraling inward. The only way it can change its state is by giving up a whole quantum of energy and jumping to a lower orbit. But if this lower orbit is already occupied by another electron, then there is nowhere for the first electron to go. It has to stay where it is, going around and around in its orbit forever. Adding Planck and Einstein's quantum to Rutherford's planetary model produced atoms that were totally stable, demonstrating, for example, why rocks—which are, of course, made out of atoms—have survived over geological time scales.

At the same time, the quantum nature of the electron's orbits would help to explain the spectrum of light that is given off by an atom. Put a pinch of sodium in a candle flame, and the electrons in each of its atoms gain energy and make quantum jumps to higher unoccupied orbits. These very energetic electrons then fall back into vacant orbits, and with each quantum jump, a photon of light-energy is released.

This theory of quantum jumps also explains why the spectral lines are so sharp. The fact that the lines in a spectrum are narrow means that their energy must be precise. But this is the whole point about a quantum of light: you cannot have something that is 99 percent or 101 percent of a quantum. Energy always comes in exactly on the nose, no more and no less. For this reason, the energy corresponding to a spectral line is precise.

Bohr was able to provide a theoretical picture of Rutherford's atom that explained not only its stability but also the sharpness and patterns of the spectrum of an atom. Beginning in 1913, Niels Bohr published a series of theoretical papers that worked out the details of his quantum atom. Scientists all over the world were excited by Bohr's ideas. His theory was easy to understand, for in essence it married the classical idea of Newtonian orbits to the new ideas of Planck and Einstein. The puzzle of the atom appeared to have been solved.

BEYOND THE BOHR ATOM

But, as time went on, a number of problems began to surface. Physicists were making more accurate measurements of atomic spectra. They discovered, for example, that under greater magnification, certain spectral lines turned out to be composed of yet finer lines. While Bohr's theory explained the main features of an atomic spectrum, physicists had to play around with theoretical variations to explain these new results. Did a proper account of these spectral fine structures require nothing more than theoretical refinements, or was something radically new required?

Between working on refinements to his theory, Bohr was now giving lecture tours. On one occasion that was to prove highly significant for the future of twentieth-century physics, he visited Göttingen, Germany, to talk about his atomic theory. It is said that Bohr was one of the world's worst speakers. When he wasn't focusing his attention on lighting his pipe, he was standing with his back to his audience and mumbling at the blackboard. David Bohm has related the story of how Bohr was once asked a difficult question. The great physicist pondered for a moment, then took out a large box of matches, which he dropped while attempting to light his pipe. Very carefully Bohr picked up the fallen matches one by one, placed them back in the box, and asked, "Any questions?"

But Bohr must have been in good form that day in Göttingen, for he excited at least one member of his audience. Indeed, he was so struck by the intensity of one of his questioners, a student, that he asked the young man to walk with him later that day. On that walk, the two scientists eagerly discussed the meaning of the atomic theory and the problems it was running into. The student's name was Werner Heisenberg, and within a few years he was destined to take the torch of knowledge from Bohr's hand and advance even further into the mysteries of the atomic world.

Heisenberg was a physics student at the University of Munich under the formidable Professor Arnold Sommerfeld. Sommerfeld was of the old school of German physics teachers, and he insisted that his students develop a deep insight into the physical world by carrying out laboratory experiments and making observations. While on a drinking trip at a nearby lake, for example, Sommerfeld directed his students' attention to the formation of waves at the back of the boat.

Sommerfeld was exactly the right teacher to give the young Heisenberg a thorough grounding in classical physics. But Heisenberg was also to meet an even more important influence on his thinking. On his first day in class, Sommerfeld introduced him to a fellow student who was to become his lifelong friend and colleague: Wolfgang Pauli.

Pauli and Heisenberg lived for physics, yet the two young men were like night and day. Pauli loved to stay out all night drinking and arguing, so that by the time he crawled out of bed the following morning, Sommerfeld's lectures were over. Heisenberg, by contrast, was an early riser, a hard worker whose favorite activities were a long walk in the mountains or sitting under a tree and reading Plato. But in spite of, or because of, their differences in temperament, the two students became deep friends, meeting each day to talk and argue about the new physics. Much later, when looking back on his professional life, Heisenberg claimed that his most important influence had not been university professors or textbooks, but his discussions with Pauli.

The two young men talked about everything—God, philosophy, the ultimate nature of matter, where the forms of nature had come from, Einstein's theory of general relativity, the new experimental results of the spectroscopists, and, of course, Bohr's remarkable theory of the atom. They made an ideal combination. Heisenberg was passionately interested in philosophy and possessed a powerful scientific imagination. Pauli, for his part, was an iconoclast. Never allowing his intellect to rest and not willing to accept any idea at face value, he was always critical, employing his exceptional physical intuition to constantly question and probe. Even a physicist as great as Einstein did not intimidate the young Pauli, who was willing to stand up in a lecture hall and criticize what the older scientist had just said. In fact, while still a student, Pauli had published a major article (today available as a book) reviewing Einstein's theory of relativity. Pauli's fearlessness and cutting criticisms soon earned him the nickname "Pauli, the whip of God."

In their talks together, Heisenberg and Pauli began to explore the whole question of Bohr's atomic theory. Was it simply a matter, as everyone thought, of modifying the theory to accept some new experimental facts—for example, in order to account for the fine structure of atomic spectra? Or was something deeper involved, some fundamental error at the heart of Bohr's idea? In science there is always the possibility that a new theory can be patched up in order to explain new experimental results. But eventually there comes a point of diminishing returns.

Pauli and Heisenberg both knew about the fall of Ptolemy's earth-centered theory of the solar system. Before the Copernican revolution, astronomers had believed that the sun and planets rotate around the earth in circular orbits. But more accurate measurements of the motion of the planets through the sky did not fit this theory. The answer at the time was to patch up the theory by adding an extra degree of complication. Instead of the planets rotating around a stationary earth in simple circles, there were now orbits within these orbits, called epicycles. And if a single set of epicycles would not work, then astronomers proposed epicycles within epicycles within epicycles. Constantly patching up the theory made it possible to explain the orbits of the planets with greater and greater accuracy. But at the same time, the theory was becoming more and more complicated and less credible. When Copernicus's new conception finally burst on the scene, scientists realized that all those complicated epicycles could be swept away simply by putting the sun at the center of the solar system. It became clear that the problem had never been one of making better refinements to Ptolemy's epicycles but had lain in the heart of the whole theory: it is the sun, not the earth, that stands at the center of the solar system.

Pauli and Heisenberg began to wonder whether something similar was the case with regard to Bohr's theory. Should physicists go on patching up Bohr's picture, or did it contain a fundamental flaw? Pauli argued that Bohr's whole theory was wrong and required a revolution as radical as that with which Copernicus had displaced the earth from the center of the solar system and replaced it with the sun. The new theory of the atom had to involve a totally new conception of reality.

Bohr's theory, Pauli asserted, had put new wine into old wineskins. It was a mismarriage of eighteenth-century physics with the latest ideas of Einstein and Planck. On the one hand, the atom was being pictured as a miniature solar system in which each electron moves in a well-defined orbit. On the other hand, Bohr had introduced the radical idea of the quantum, in which electrons execute mysterious "quantum jumps" between orbits.

Bohr talked about "orbits" in which electrons move around on paths just like stones and cannonballs. In other words, electrons were being treated as very tiny but quite traditional objects, smaller versions of things that would be quite familiar to Newton. Yet these same electrons underwent mysterious quantum jumps as they moved from one orbit to another.

To talk about planetary orbits was to employ the sort of language that physicists had used since the time of Newton. But Pauli believed that the quantum principle was pointing toward something revolutionary. Bohr's theory was not radical enough; it was still clinging to familiar "classical" ideas.

Pauli had met the great Einstein and talked to him about how he had created the theory of relativity. Einstein had explained that it was necessary to go deeply into the whole meaning of time and of space. The deepest questions are philosophical in nature, and it was out of this deep thought that the new theory had been born. Einstein had also said that a good theory tells a physicist what the world must look like. What can be known about the world in fact emerges out of the theory itself. This idea that the theory should talk about the "observables" of nature was to play a key role in Heisenberg's thinking and is central to what has become the standard, or so-called "Copenhagen" interpretation, of quantum theory.

HEISENBERG'S REVOLUTION

A year after his meeting with Bohr, Heisenberg had moved to Göttingen, where he was working as physicist Max Born's assistant and grappling with the fundamental problems of Bohr's theory. Over the next two years, Heisenberg continued to correspond with Pauli and to puzzle over just what a radically new theory would mean. The desire for some totally revolutionary way of thinking about the quantum world became an obsession with Heisenberg. In June of 1925, suffering from a severe attack of hay fever and worn down by his failure to make a breakthrough in the theory of the atom, Heisenberg decided to take a vacation on the grassless island of Helgoland in the North Sea.

Away from Göttingen and all his worries about the physics of the atom, Heisenberg gave himself over to relaxation and the pure physical exertion of walking and swimming that he so much enjoyed. For a time he could put aside the burning anxiety of his ideas. Then one evening everything suddenly fell into place. The twenty-four-year-old physicist had a feeling of vertigo, as if he were looking into a powerful microscope and watching the very processes that happen within the atom. "I was deeply alarmed," Heisenberg was later to write. "I had the feeling that, through the surface of atomic phenomena, I was looking at a strangely beautiful interior, and felt almost giddy."

In a burst of inspiration, he wrote down equations as they occurred to him and quickly began to make some calculations. Yes, everything seemed to be working out. The theory was perfect; it accounted for everything and in a particularly neat way. From now on, there would be no more talk of orbits, no electrons behaving like microscopic planets. The old-fashioned "classical reality" had been thrown away, and in its place stood an austere mathematical expression of a world that lies beyond anything we will ever touch or see. Quantum theory had been born. "I was far too excited to sleep," Heisenberg wrote, "and so, as a new day dawned, I made for the southern tip of the island, where I had been longing to climb a rock jutting out into the sea. I now did so without too much trouble, and waited for the sun to rise."

Hurrying back to the mainland, Heisenberg met Pauli in Hamburg and told him about his discovery. Then, in Göttingen, Heisenberg discussed his theory with Max Born and another leading physicist, Pascual Jordan. A matter of days later, Heisenberg had written down his ideas in a formal scientific paper, which he sent for review to Pauli. Max Born later recalled Heisenberg giving him a copy of the paper and asking if it was worth publishing. Born was tired and did not bother with it for a few days. But as soon as he picked it up, he was fascinated by the audacity of what Heisenberg had done. At once he wrote to Einstein and told him the good news.

Pauli read the paper and within days was using the new theory to make calculations of the spectrum of the hydrogen atom. The results were to be one of the first great triumphs of the quantum theory. Meanwhile Heisenberg, along with Born and Jordan, was busy unfolding the theory's theoretical implications. By the end of the summer, Heisenberg was in Copenhagen talking to Bohr.

Within a matter of months, the world's leading physicists were discovering just how radical was this new quantum theory of Heisenberg's. While Bohr's theory of the solar-system atom had hung onto the familiar ideas of electron orbits, Heisenberg's theory was totally revolutionary. Not only was the quantum theory probabilistic, but it involved the famous Heisenberg uncertainty principle, which shattered physicists' intuitions about how the world really looks. The controversy over quantum reality had been born.

Matrices and Uncertainty

What exactly had Heisenberg done that so amazed the physics community? His discussions with Pauli had led him to the point where he was forced to cut himself free from all familiar pictures and appeals to "classical ideas" of reality. The theory of the quantum world had to be fresh and new; there was no room for outmoded concepts.

But did this mean that the baby had to be thrown out with the bathwater? If the familiar ideas of physics were all abandoned, then what would be the starting point? At least Heisenberg could work with the atomic spectra, that pattern of frequencies that characterizes each individual atom. Einstein had pointed out that a good theory tells us what is observable in nature. What if these patterns of light frequency were the observables of the new quantum theory?

Assuming that these numbers—these energy frequencies measured by the spectroscopists—had to be the basic data of the theory, and that they were somehow related to energies within the atom, Heisenberg used his intuition to arrange them into patterns. The patterns looked something like the array shown here.

FIGURE 2-5

$$
\begin{matrix}
a_{11} & a_{12} & a_{13} & a_{14} & a_{15} \\
a_{21} & a_{22} & a_{23} & a_{24} & a_{25} \\
a_{31} & a_{32} & a_{33} & a_{34} & a_{35} \\
a_{41} & a_{42} & a_{43} & a_{44} & a_{45} \\
a_{51} & a_{52} & a_{53} & a_{54} & a_{55}
\end{matrix}
$$

A matrix is an array or pattern of numbers. Here the symbols a_{nm} stand for these numbers. Thus, a_{nm} is the number in the *n*th row and *m*th column. Physicists and mathematicians are particularly interested in matrices with particular symmetries and patterns, for example, matrices in which a_{nm} is always the same as a_{mn}, or in which the numbers that do not lie on the diagonal a_{nn} are all zero.

It became clear that there was a special rule for multiplying and working with these patterns. Already the logic of the quantum world was coming into focus, and it lay in the mathematical rules for using the number patterns. Heisenberg tried out his idea on one of the simplest systems known to physics—the quantum analogue of a simple, swinging pendulum. A pendulum swings from side to side; the quantum analogue is a vibrating atom that is radiating light. Sure enough, the system could be defined by two square patterns of numbers—these numbers corresponding to experimentally measurable properties of the atom.

The next step was to know how to work with these number patterns; for example, how to multiply them. This new multiplication rule was one of the insights that occurred to Heisenberg on Helgoland. Suddenly everything had fallen into place, for the patterns and numbers and the rules for multiplying them *were* the basis of the new quantum theory. As we shall see, from these simple rules and patterns would follow everything—the spectra of atoms, the uncertainty principle, and all the subtle arguments of quantum reality that were to come.

As we have seen, on his return from Helgoland, Heisenberg quickly wrote down his equations and showed the draft to Born. When Born finally got around to reading Heisenberg's paper, he was so fascinated that he could hardly sleep at night. The more he looked at Heisenberg's patterns of numbers and the way they multiply together, the more he sensed that something familiar lay behind them. Then one morning, he suddenly saw the light. These patterns are what mathematicians call matrices, something that had been discovered in a totally different context over fifty years earlier by the eccentric Irish mathematician William Rowan Hamilton.

By writing Heisenberg's patterns in a different way, Born could draw upon a powerful mathematical tool known as matrix algebra. He roped in his student Pascual Jordan to help, and soon Born and Jordan—and then Born, Heisenberg, and Jordan—were writing about the new matrix mechanics, as Heisenberg's quantum theory became known. In England a brilliant young student, Paul Dirac, was also figuring out the implications of this theory of quantum matrices.

The individual numbers in these matrices correspond to the results of a measurement made on the atom. Moving these matrices around and manipulating them mathematically therefore corresponds, in a theoretical way, to what happens when physicists actually make a measurement or carry out an experiment. Setting up a piece of apparatus to observe the atom, then recording a click on a Geiger counter or the movement of a meter is theoretically equivalent to working with a mathematical matrix and selecting one of the numbers it contains. In other words, making experimental measurements could now be described in terms of working with Heisenberg's matrices. The results of any measurement correspond to picking out one of the numbers in the matrix pattern.

On the surface, things don't seem too complicated, but it is exactly at this point that quantum theory parts company with the physics of the everyday world. In discovering that the quantum world must be represented by matrices, Heisenberg had introduced a curious new element into physics. When you multiply two numbers, it does not matter in what order you do it: 2×4 is exactly the same as 4×2. The answer is always 8. Likewise, 5×9 is the same as 9×5. This property, that the order in which numbers are multiplied does not matter, is called commutation. Ordinary numbers commute, so $(4 \times 5) - (5 \times 4)$ is always zero.

FIGURE 2-6

$$q = \begin{matrix} 0 & q_{01} & 0 & 0 & 0 & \cdots \\ q_{10} & 0 & q_{12} & 0 & 0 & \cdots \\ 0 & q_{21} & 0 & q_{23} & 0 & \cdots \\ \vdots & \vdots & \vdots & \vdots & \vdots & \cdots \end{matrix}$$

One of the matrices used by Heisenberg and taken from a paper by Max Born and Pascual Jordan, written in 1925. Heisenberg was attempting to calculate the motion of an oscillating quantum particle. His calculations were complicated and cumbersome until Born and Jordan showed how they could be put into matrix form.

Matrices, however, are multiplied in a different way from ordinary numbers, so they do not always commute. If A and B are both matrices, then $A \times B$ may give a different answer from $B \times A$. To put it another way, $(A \times B) - (B \times A)$ may not always be zero. Matrices that multiply in this way do not commute. When you are dealing with numbers, ordinarily it does not really matter in what order you carry out the operations. With matrices it does matter, since a different answer can result.

Take a simple example. Go two steps forward, then turn to the side. You end up in a different place than if you had first turned to the side and then taken two steps. Similarly, putting on your socks and shoes is quite different from putting on your shoes and then your socks. With operations that do not commute, you have to take care to do them in the right order.

In his moment of insight on Helgoland, Heisenberg had discovered a new form of physics in which the basic entities of reality are represented by matrices, which do not always commute. The implications of this discovery are profound. Heisenberg's matrices, or patterns of numbers, correspond to measurements that can be made on an atom or other quantum particle, and this suggests that the order in which quantum experiments are carried out could be crucially important! Suppose, for example, that one matrix or pattern of numbers, called P, corresponds to measuring position, and the matrix S corresponds to measuring speed. The rules that Heisenberg had discovered tell us that $P \times S$ is not equal to $S \times P$. The two matrices do not commute, and the experimental results must therefore depend on the order in which the measurements are carried out!

This conclusion is so important that it deserves spelling out in detail. Suppose that an experiment is set up to locate the position of a particle, and this is followed by a measurement of its speed. Two well-defined results are obtained. But suppose now that the speed is measured first; when we come to measure the particle's position, we discover a totally different result! It is as if one measurement has disturbed or interfered with the other—as if the order in which measurements are carried out actually affects the result. This was one of the most surprising results of the quantum theory, yet it follows quite logically once matrices are chosen as the natural language for quantum physics.

The Cambridge physicist Paul Dirac was also worrying about the noncommutation of Heisenberg's matrices. In fact, a friend from Cambridge told me the following presumably apocryphal story about Dirac. Apparently Dirac's wife had advertised the sale of a dresser. A respectful student telephoned Dirac's home, and the professor answered.

"Hello?"

"Do you have a dresser for sale, Professor Dirac?"

"What? Oh . . . er . . . yes."

"Well . . . how long is it, please?"

"Wait a moment."

There is a long pause while Dirac shuffles off to measure the length of the dresser. Eventually he returns and says, "It appears to me to be six feet long."

"And, if you don't mind, how tall is it, Professor Dirac?"

This time there is an even longer pause before Dirac comes back to the telephone and asks, "Do they commute?"

In asking if the height and the length of the dresser commute, Dirac was worrying that by moving the tape to measure height instead of length, he would actually be changing his first result. In the Alice-in-Wonderland world of quantum physics, after measuring the height of the dresser at four feet, Dirac may then discover that its length is no longer six feet but ten feet or even two feet!

Of course, things don't work that way in our large-scale world. It does not matter in what order we measure velocities or distances or the sizes of dressers; they all work out the same in the end. But in the world of atoms, things are very different, and that is what led Heisenberg to propose his famous uncertainty principle.

The Uncertainty Principle

Suppose you stand on the bathroom scale in the morning to find your weight. You may not always like the answer you get, but you know that, unless someone has been tampering with the scale, the answer is good to a pound. You then measure your height, which is accurate to around one-tenth of an inch.

There are always errors when you make a measurement—errors in the bathroom scale or the length of the tape measure. But one thing you do know for certain is that you actually do *have* a definite weight and, by buying a better scale, you can find that weight more accurately. You can even go to the trouble of taking the scale to a laboratory to be tested and working out tiny corrections to fractions of an ounce. Through greater and greater effort, you get an answer that comes closer and closer to your true weight.

But what would happen if, after measuring your height, you found your weight had changed? What would happen if, each time you measured your height, your bathroom scale gave you a different weight? Would this mean that you did not really have a single well-defined weight at all, that your weight was not a constant thing but kept changing?

Luckily for our sanity, things don't work that way in our large-scale world. We know that at any given moment we have a definite height and a definite weight, and that measuring one has no effect on the other. But in the quantum world, things are very different, for the result of making one measurement does affect the other. Thus, it does not matter how refined the measurements are or how carefully they are made, there will always be uncertainty in the final results.

Try measuring the velocity of an electron. The more accurately this is done, the more uncertain becomes the electron's position. Pin down the position of the electron, and its velocity becomes uncertain. This is Heisenberg's famous uncertainty principle. It follows directly from that curious rule for working with patterns of numbers that he had discovered while on Helgoland.

Heisenberg was even able to give the actual size of the uncertainty involved. It turns out to be a number called Planck's constant—a number first derived by Max Planck in order to explain the size of a quantum of energy:

$$\text{Error in position} \times \text{Error in velocity} \approx \text{Planck's constant}$$

Decrease the error in position, and you increase the uncertainty of the velocity. Attempt to pin down velocity, and position becomes uncertain.

In fact, when you think about it, Heisenberg's uncertainty principle makes perfect sense. A measurement of, say, the position or velocity of an electron involves an interaction with experimental measuring apparatus, and any interaction implies some interchange of energy. After all, if no energy is interchanged, then nothing would have changed in the measuring apparatus. Planck and Einstein had pointed out that this interaction must involve at least one quantum of energy. Since there can never be a fraction of a quantum, the minimum uncertainty in a measurement must be the size of one quantum. This size is given by Planck's constant.

A quantum of energy is very small indeed—so small that it cannot be detected by our normal scale of things. This is why we do not have to bother about Heisenberg's uncertainty principle when we weigh ourselves or have the length of a cottage lot surveyed. But at the scale of atoms and electrons, Planck's constant becomes important and can dominate everything.

Heisenberg's Microscope Experiment

In an effort to understand the implications of his uncertainty principle, Heisenberg came up with his famous microscope experiment. The uncertainty principle follows directly from the equations that Heisenberg had written down, but it still takes some believing. What was needed was a simple illustration that would explain how quantum uncertainty works in practice. To do this, Heisenberg devised a hypothetical device, or "thought experiment," that shows why the uncertainty principle is needed. However, its deeper meaning was to transform yet again what we mean by quantum reality.

When astronomers want to plot the future path of a new comet, they must first measure its present position and its speed and direction of travel. Knowing the speed by itself is not enough, for it is important to know the direction in which the comet is heading. The combined values of speed and direction are called velocity. A car may be traveling along the Brooklyn Bridge at 30 mph, but it's also important to know its direction of travel, that is, whether the car is heading into Manhattan or toward the suburbs. So scientists tend to talk about velocity rather than speed.

Given these two facts, position and velocity at a single

instant of time, it is possible to predict the position of the comet at any future time. The position and velocity at some point are the key factors in determining the path of any comet, missile, pool ball, or baseball, and consequently these two variables play a fundamental role in classical physics. However, if information about one of them is uncertain, then it is no longer possible to pin down a path and make predictions.*

These are exactly the same quantities that do not commute in quantum theory. Position and velocity are represented by noncommuting matrices, or noncommuting patterns of numbers. This means that they can never be simultaneously measured without some uncertainty creeping into the answer. If one is known, the other becomes uncertain. But this means that the path of an atom or electron can never be pinned down.

Heisenberg set out to demonstrate why this must always be the case. His argument goes the following way: Suppose you try to find the exact position of a moving electron. One way of doing this is to use a microscope. With ordinary visible light, the wavelengths involved are much larger than the electron itself, and any measured position would be vague. The answer is to use light of extremely short wavelength such as high-frequency gamma rays. With the help of a gamma ray, it becomes possible to locate the position of the electron more accurately. But the problem is that a gamma ray has high energy. In hitting the electron, this gamma ray quantum makes an impact that changes the electron's velocity. Any attempt to pin down the position of the electron makes its velocity uncertain.

The way to reduce the uncertainty in velocity is to use light

---

*Physicists more generally speak of position and *momentum* rather than position and velocity. Momentum is the actual variable that appears in the important equations of classical and quantum physics. Momentum is simply calculated by multiplying the velocity of an object by its mass. (Where particles interact in a complicated way, and when we deal with photons of light, it is necessary to modify this simple definition.) To avoid too much technical language, I will continue to use the term *velocity*, and readers with a technical background will know that I mean momentum.

of a longer wavelength (and lower energy), which will have much less effect on the electron's velocity. But using light of longer wavelength means increasing the uncertainty in the electron's position. The more precisely you measure position, the more uncertain becomes velocity. The more accurately you determine velocity, the more uncertain is position.

Heisenberg's conclusion is clear: what you gain in one area, you lose in another. The universe conspires against us and prevents us from ever knowing the exact combination of position and velocity that defines the path of an electron. And if we can never know the paths of electrons, protons, and atoms, then the quantum world is forever uncertain, and, as French philosopher Bernard d'Espagnat observed, its reality is veiled from us.

This veiled reality came as a great shock to many physicists. Pauli's intuition had been correct. When you enter the quantum world, everything changes, and there is no longer any possibility of certainty.

Heisenberg wrote out his account of the hypothetical microscope experiment and traveled to Copenhagen to get Bohr's reaction. The response was a tremendous shock to the young physicist. Indeed, by the end of that day, Heisenberg was in tears, for Bohr argued that while the uncertainty principle was correct, Heisenberg had missed an essential point in his microscope experiment. The true meaning of quantum uncertainty went much deeper than Heisenberg had ever imagined.

A Pathless Land

In his microscope experiment, Heisenberg had shown how every attempt to measure a property at the quantum level results in an interaction with the measuring apparatus itself. Since at least one quantum of energy must be exchanged every time a measurement is involved, this interaction can never be reduced or eliminated—for quanta are indivisible. All this was perfectly correct, Bohr agreed. But Heisenberg had then made a fundamental mistake.

In setting out his argument, Heisenberg had assumed that an electron does in fact have an actual path but one that we can never know exactly. At each instant of time, Heisenberg had assumed an electron has a well-defined position and is moving with a particular velocity. As it travels across the laboratory, for example, the electron moves in a given path. Any attempt to determine one of the two parameters of this path, position or velocity, will interfere with the other. Nature is preventing us from knowing about the exact motion of the electron. Its precise path is veiled from us.

But that's exactly where you made your mistake, Bohr said. Your basic assumption is wrong: *the electron does not have a path!* In fact, it's not even correct to talk about the electron "having" both a position and a velocity. What the uncertainty principle is trying to tell us, Bohr argued, is that the very concept of a path is ambiguous at the quantum scale of things.

Just as Bohr himself had once tried to hold onto the notion of planetary orbits when he created his quantum picture of the atom, so now Heisenberg was trying to retain the idea of paths and well-defined (if experimentally uncertain) positions and velocities for the electron. These are classical ideas, Bohr said, and they will have to go. The uncertainty principle is really pointing to the basic ambiguity of the quantum world. We have to let go of all our familiar and comfortable ideas. It is not so much that quantum interactions forever prevent us from knowing the exact path of an electron, but rather the very *notion* of path is ambiguous in quantum theory. The electron's path is not what is uncertain; it is the concept of "path" that is inappropriate!

Even d'Espagnat's description of the quantum world as a "veiled reality" does not correspond to the deeper point Bohr was making. A reality that is forever veiled from us implies some sort of deception on the part of nature that prevents us ever touching the true essence that lies beyond appearances. But it is not so much that quantum reality is veiled from us as it is that the very notion of an independent reality no longer applies. The whole meaning of quantum reality had become profoundly different from anything physics had experienced before. Perhaps the very idea of reality no longer applied at the quantum level.

COMPLEMENTARITY

In rejecting the attempt of physics to hold on to the last shreds of a reality in which objects possess well-defined properties, Bohr introduced a new notion, which he called *complementarity.* Complementarity, Bohr believed, was a deep new principle, deeper even than the uncertainty principle. Complementarity means that the quantum universe cannot be contained within a single description. Rather, complementary and even paradoxical descriptions are required—like wave and particle. The closer one focuses on one description, the more ambiguous the other becomes.

At first this new idea came as a great shock to Heisenberg, but in the end he was willing to accept Bohr's dramatic new idea. Along with Pauli and Heisenberg, Bohr argued that quantum theory must cut itself free from all the outmoded concepts of classical physics. The world of the atom is profoundly different. Indeed, the whole issue of quantum reality was dissolving into quantum ambiguity. Was there a well-defined but hidden reality within the atom? Or must this very notion of reality be abandoned, as Bohr suggested, and replaced by a collection of complementary and paradoxical descriptions?

Working together in Copenhagen, the trio struggled to make sense of the seemingly paradoxical implications of the quantum theory. Already there were rumors that scientists of the stature of Einstein were not happy with this wholehearted rejection of the familiar notions of reality. Admittedly Einstein had turned space and time on their heads, but the great physicist still believed that there was a tangible, objective, independent reality out there and that nature was founded on rational laws. Now all this seemed to be disappearing into uncertainty, ambiguity, probability, and complementarity.

Then, just as things were coming to a head, a German physicist named Erwin Schrödinger came up with a rival theory, one that seemed to bring everyday reality right back into the world of the atom.

3
AN INDIVISIBLE WHOLENESS

Schrödinger's alternative theory of the quantum world, called wave mechanics, put the Copenhagen group into a state of shock. For several months Bohr, Heisenberg, and Pauli had been working on what they believed to be the definitive theory of the quantum world, one that rejected classical reality for ambiguity and complementarity. Now Schrödinger had come up with a totally different approach, one that appeared able to account for all the quantum features of nature without becoming embroiled in abstruse discussions about the ultimate nature of reality.

Schrödinger's wave theory was based upon the ingenious idea of Louis de Broglie's that matter has a wavelike nature. Although de Broglie's idea did not at first attract much attention, a copy of his thesis was sent to Einstein by Paul Langevin, a close friend of Marie Curie's. Einstein immediately realized the importance of what de Broglie had done. In fact, he seems to have been thinking along similar lines himself. Both men agreed that electrons must have a wavelike nature in addition to their particle nature.

The new ideas of de Broglie and Einstein quickly came to the notice of Erwin Schrödinger, a German physicist who was then working at the University of Zurich. Schrödinger was thirty-nine when he created his wave mechanics, a remarkably late age for a theoretical physicist to have made his first important breakthrough. The Viennese-born physicist had made a series of steady contributions to science but nothing out of the ordinary. Indeed, some eight years earlier, he had even thought about giving up physics.

Now Schrödinger was beginning to think about de Broglie's thesis. If electrons truly act like waves, he argued, then why not work out a proper wave description for them? Just as sound waves in a flute or organ pipe can be described by mathematical equations, so too must the electron waves around the nucleus have an associated wave equation.

FIGURE 3-1

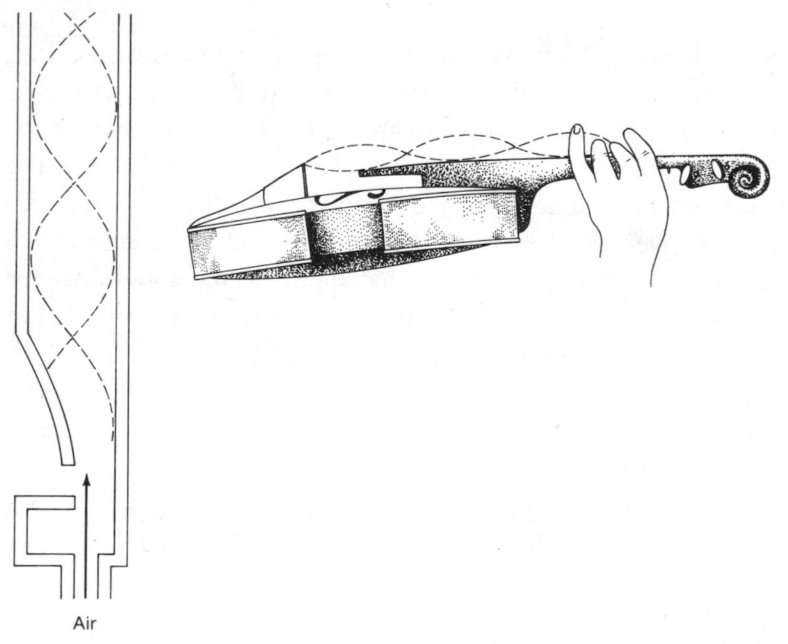

The characteristic notes of musical instruments are the result of standing waves. A certain number of standing waves fit easily into an organ pipe, or along a violin string.

Physicists had already noticed that the spectrum of the hydrogen atom, which is related to its internal energy states, is a little like the pattern of notes on a musical scale. Could that be an additional clue to the wavelike nature of the atom?

Musical notes in an organ pipe are created by what are called standing waves. These are waves that fit exactly into the space inside the pipe and carry all the energy associated with a particular note. Other waves, called transients, rapidly die away and are not heard.

Standing waves of several different wavelengths can fit into a given pipe. The wavelength determines the pitch of the note produced by the pipe. The more wavelengths of a standing wave, the higher is the note. Thus, a whole series of different notes can be produced from various standing waves in a single pipe. The pattern of notes therefore depends on which waves can fit into the pipe.

Standing waves in flutes, trombones, oboes, horns, and trumpets generate patterns of notes. Likewise, a violin, cello, and guitar use standing waves. The ends of each string are fixed, and the string vibrates at its natural, lowest frequency. Place a finger

FIGURE 3-2

(a) The lowest note of a vibrating string is produced when a single wave fits into the string's length.

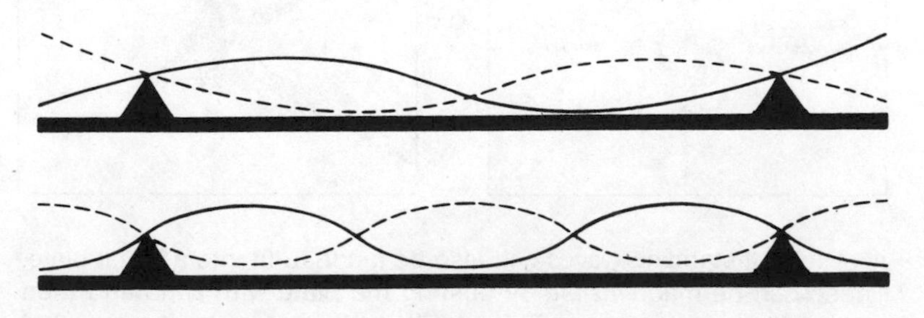

(b) Two or more waves can be fitted to produce a higher note.

midway along the string, and you can sound a note that is an octave higher. In each case, the energy given to the string by plucking it or drawing a bow along it is transferred to a standing wave. Any other sort of vibration—one that does not fit exactly into the length of the string—will rapidly die away. In other words, energy can only be stored within the standing waves that fit exactly into the length of the violin string or organ pipe.

FIGURE 3–3

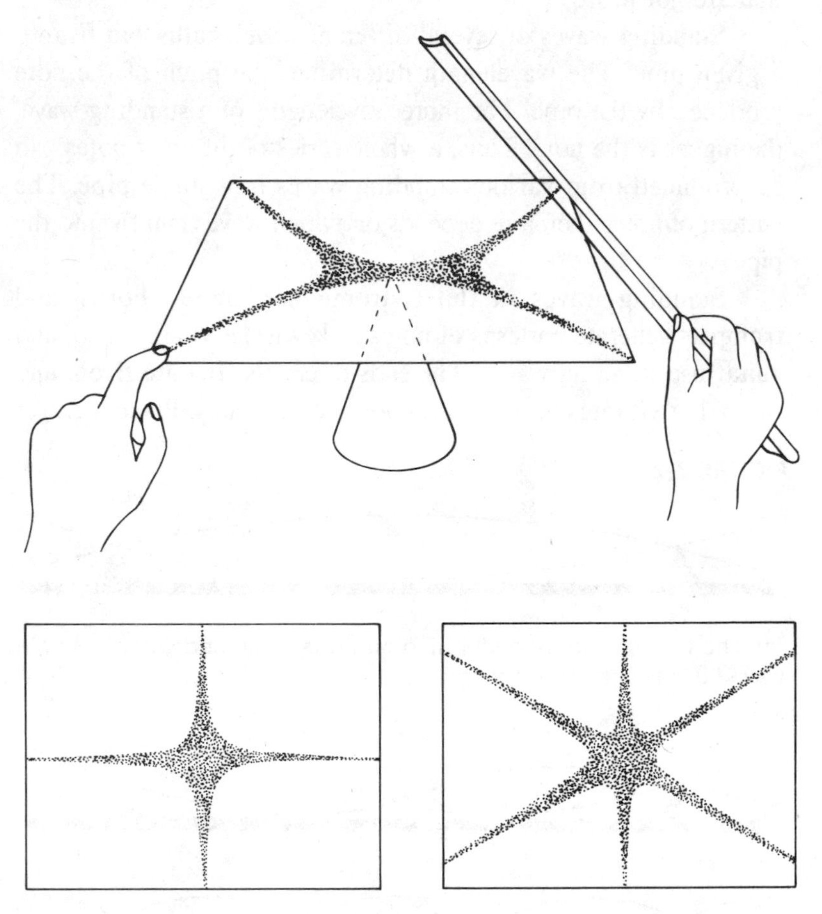

A variety of standing waves can also be made to fit into a metal plate. These can be made visible by dusting the plate with sand and then making it vibrate with a violin bow. (The different patterns are called Chlandi's figures.)

Could there, Schrödinger asked, exist standing electron waves around the nucleus of a hydrogen atom as well? If true, the electron must have a fixed set of energy levels, like the notes of a violin or organ pipe. The fixed energy levels of each atom, Schrödinger assumed, are associated with different patterns of standing waves. Standing waves are stable; moreover they have precise patterns of energy. The following illustration shows standing-wave patterns that fit exactly around an atom. Any waves that do not fit perfectly would rapidly die away. In a single step, Schrödinger was able to explain all the known facts about the atom, its stability, and spectrum without any of the curious assumptions made by Bohr and Heisenberg.

FIGURE 3–4

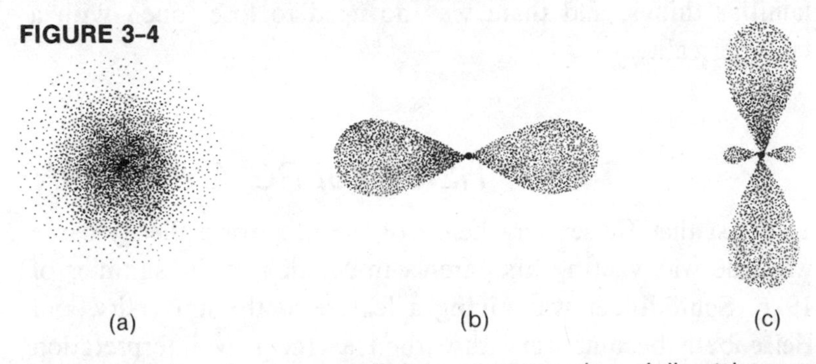

Just as vibrational waves can fit into an organ pipe, violin string, or metal plate, so too can standing waves fit around an atom. This illustration shows (a) the lowest oscillation—called a 1s state, (b) p-states, and (c) d-states.

But would the pattern of these energy levels be exactly the same as that calculated by Pauli using Heisenberg's matrix quantum theory? Would it agree with experiment? Schrödinger quickly worked out the mathematics of the electron waves, using a wave equation that is now called Schrödinger's equation. The equation turned out to explain the hydrogen atom perfectly. There was now a natural and realistic explanation of why the atom has a discrete series of energy levels—they are the only ones that correspond to standing-wave patterns. Similarly, the jumps between the various standing waves correspond to jumps in energy and reproduce exactly the spectrum of the hydrogen atom.

Using de Broglie's matter waves, the stability and spectrum of hydrogen could easily be explained without the need for all that metaphysics about a veiled reality. Schrödinger's paper on wave mechanics was published in January 1926. Those few months since June 1925, when Heisenberg's first insight had appeared, were one of the most important periods in all of physics.

Just six months after Heisenberg's original breakthrough in quantum theory, it almost looked as if the matrix theory had been nothing more than a passing phase. Now the whole thing could be done much more simply with Schrödinger's wave mechanics. In addition, waves, even if they happened to be matter waves, were familiar things, and there was no need to lose touch with a tangible reality.

HEISENBERG AND BORN

The first that Heisenberg heard of the new wave theory came when he was visiting his parents in Munich in the summer of 1926. Schrödinger was giving a lecture at the university, and Heisenberg became very disturbed as the new interpretation unfolded. He began to make objections, asserting that the whole approach seemed confused. Yet Heisenberg's remarks were swept away by one of the senior professors, who said that it was time to put an end to "atomic mysticism." But could Schrödinger be correct? Was the ultimate nature of the quantum world that simple?

In Göttingen Max Born was also taking a hard look at Schrödinger's equation. Schrödinger had applied his theory to the hydrogen atom, in which a single electron is outside the central nucleus. The equation certainly looked correct. It described waves in three-dimensional standing patterns around the nucleus. In addition, the energy associated with each pattern fully accounted for the hydrogen spectrum. But what happened when Schrödinger's equation was generalized to helium, with two electrons, or lithium, with three? Would it still work?

At this point, the realistic interpretation proposed by Schrödinger began to dissolve away, to be replaced by something more subtle and difficult to understand. Born discovered that the wave equation for a helium atom with two electrons could not be written in our familiar three-dimensional space but had to be expressed in a space of six dimensions. Rather than two electron waves moving in three-dimensional space, the wave equation seemed to be referring to a single wave in some mysterious six-dimensional space. What was worse, in the case of lithium, a space of nine dimensions was needed. Every time an electron was added to the scheme, an additional three dimensions were required!

What were real, physical waves doing in an abstract multidimensional space? Things were no longer making sense.

Max Born puzzled over the meaning of the wave theory until suddenly the realization hit him that the beautiful pictures Schrödinger was producing were not of physical matter waves at all. Certainly an electron has a dual particlelike and wavelike nature, but Schrödinger's equation could not be about actual material vibrations—physical waves—since these waves do not exist in our physical three-dimensional space. They had to be written in a multidimensional space.*

The wavelike patterns produced by the equation had nothing to do with material waves at all, Born perceived. Schrödinger had been fooled. These were not electron waves but patterns of probability. Quantum theory is essentially probabilistic, and the solutions to Schrödinger's equation turn out to be pictorial expressions of these probabilities. Once again, reality had slipped away to be replaced by something ambiguous and subtle. Schrödinger's wave theory had not done away with atomic mysticism after all; it had only deepened its effect!

---

*Today, with superstring theories set in ten-dimensional spaces, the idea of leaving our familiar three spatial dimensions may not appear that unusual. But to the quantum pioneers of the early decades of this century, this had become a major strike against Schrödinger's theory. After all, the number of dimensions went up with the number of electrons involved. In the case of a uranium atom, a 276-dimensional space would be needed!

WAVES AND PROBABILITIES

The idea that quantum theory is about patterns of probabilities may take some getting used to. It was something that Einstein never could accept, proclaiming often that "God does not play dice with the universe."

Probabilities are familiar in everyday life. The chance of getting heads in a coin toss is 50 percent. There may be a 10 percent chance of rain today. Insurance companies publish tables for term life insurance that are based on probabilities of people of a given age dying within a given term. Given such a table, all you have to do is select your age and look up the corresponding cost of one year's insurance. These tables tell you that the older you are, the more it costs to insure your life for the next year. Healthy young men or women in their early twenties have little chance of dying in the following year. But when they reach their sixties, the chances of dying in the next year increase. For this reason, it costs more to purchase a year of term insurance as you get older.

Of course, the insurance company has no idea exactly when *you* are going to die. You may live for the next thirty-five years and die in your ninety-ninth year—or you may get hit by a truck tomorrow. Life expectancy tables cannot predict the fate of one individual. But they can do the following: Take 1,000 fifty-year-old men. The insurance company knows from its long experience of insuring people that it is a good bet that x number of these men will die within the next year. While a single event—the fate of a single individual—lies beyond their ability to predict, the insurance tables give the average outcome over a large number of individuals with confidence. Using these tables, insurance companies can predict that if they insure the lives of a large number of fifty-year-old men, they are not going to lose money over the following year.

Insurance tables can be thought of as a pattern of probabilities given by age. Corresponding to each particular age, there is a given probability that an individual will die within the next twelve months. Just as insurance tables give probabilities in time,

FIGURE 3–5

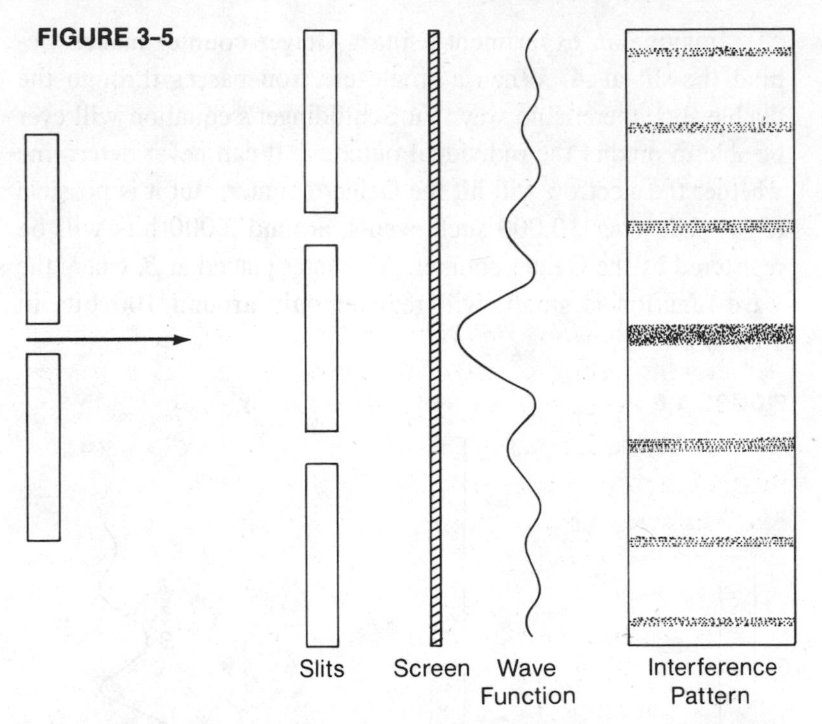

The wave function behind a double slit. The size of the wave function corresponds to the probability of discovering an electron in that particular region of space.

so Born was suggesting that Schrödinger's equation can be used to calculate a pattern of probabilities in space. The mathematical solution to Schrödinger's equation is called the wave function, and it is related to the chance of finding an electron within a particular region of space. Where the wave function is large, there is a good chance of discovering an electron; where it is small, the probability of discovering an electron will be very low. The wave function can never predict the actual position of an electron. Rather it gives us the probability of scoring a hit if we locate a Geiger counter at a particular position in space.

The above diagram plots the shape of the wave function behind a double slit. Note that its fluctuations in size correspond exactly to the pattern of light and dark bands in the interference pattern.

Imagine an experiment using a Geiger counter placed behind the slit at *A*. When a single electron passes through the double slits, there is no way that Schrödinger's equation will ever be able to predict the individual outcome. It can never determine whether the electron will hit the Geiger counter. But it is possible to say that, over 10,000 such events, around 2,000 hits will be registered by the Geiger counter. A counter placed at *B*, where the wave function is small, will register only around 100 hits in 10,000 events.

FIGURE 3-6

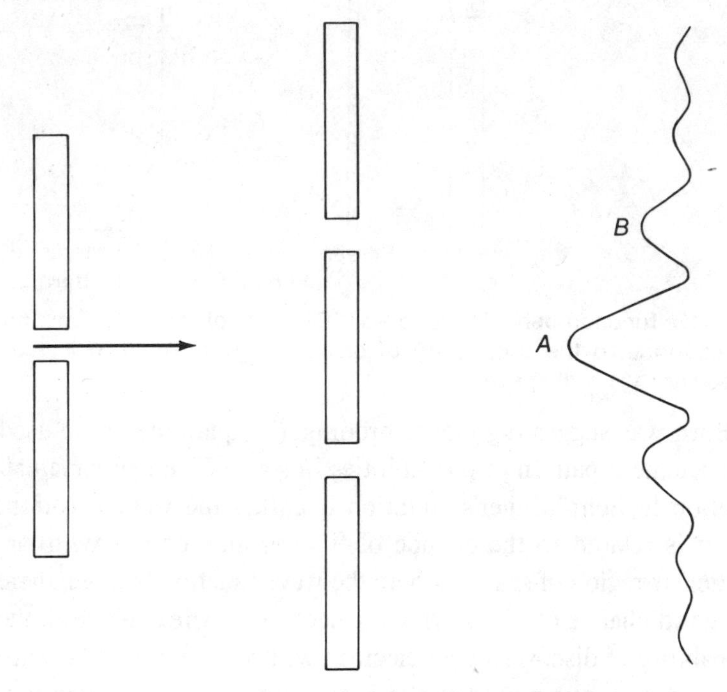

A Geiger counter placed at *A* has a high probability of registering after an electron has passed through the double slits, while at *B*, the chance of the counter responding is very low.

In this example, while an individual event is totally indeterministic, it is possible, over a very large number of occurrences, to give good probabilities. It is like betting on a poker hand. You have no way of telling how your hand will improve on the draw, but you can calculate the odds on a straight flush or a full house. This is exactly analogous to what Schrödinger's equation is doing. Instead of describing physical matter waves, it is describing the probabilities of finding an electron at a given location in space. The larger the wave function, the greater the probability of finding the electron.

In Niels Bohr's early theory of the atom, each electron moved in a definite orbit. In the wave theory of Schrödinger, it is never possible to pin down the electron. Rather, the equation suggests that the electron has the greatest chance of being found in those regions where the wave function is largest. The diagrams on page 51 plot this wave function, or probability pattern, for several energy levels of the hydrogen atom.

Born was soon able to show that, in terms of its predictions, Schrödinger's approach is essentially similar to that of Heisenberg's. One deals with matrices, related to observations, the other with patterns of probability, but in the end, they come up with identical answers. Any appeal to "classical reality" had been rejected forever.

Schrödinger was at first reluctant to accept the conclusion that his wave function did not correspond to anything tangible but simply to patterns of probability. Niels Bohr, aware of Max Born's theoretical investigations, summoned Schrödinger to Copenhagen to give a series of lectures. Once and for all, the myth that atomic nature can be pictured in classical terms, be they waves or particles, had to be put to rest. Bohr argued with Schrödinger to a point that Heisenberg later described as fanatical. Each time Schrödinger pressed his theory forward, Bohr would refute it and grind down the opposition even further.

Bohr was trying to force Schrödinger to accept that quantum theory had made a complete break with classical reality. Physics could never go back to old, familiar ideas of orbits, paths, and wave patterns. But Schrödinger would not agree and became so ill that he took to his bed. Even then Bohr would not leave him alone—the indefatigable Dane sat on Schrödinger's bed and urged him to accept the interpretation that was being worked out in Copenhagen. In the end, totally exhausted, Schrödinger cried out, "I'm sorry that I ever started to work on atomic theory," and gave in.

A few days later, after Schrödinger had returned home, Bohr received a letter from him. Schrödinger had gone back on his recantation and, for the rest of his life, felt uneasy with the Copenhagen interpretation and its rejection of all the familiar props of "classical reality."

Schrödinger's and Heisenberg's two approaches were, at a deeper level, equivalent. While one dealt with probability patterns and the other manipulated matrices or patterns of numbers, when it came to actual calculations, the two approaches produced identical results. In fact, it turned out to be something of an advantage to have two ways of looking at one problem. In dealing with some systems, the Schrödinger approach proved easiest, while for others the Heisenberg matrices worked best.*

THE COPENHAGEN INTERPRETATION

The objections of physicists of the caliber of Einstein, Planck, and Schrödinger to the new ideas that were emerging from Copenhagen forced Bohr and his two colleagues to create what philosophers call an epistemology—a theory of knowledge. In this case, it was a theory of what can be known about the atomic world. This theory has become known as the Copenhagen interpretation of quantum theory.

Philosophers have always been concerned with questions like "How can we know anything for certain?" and "What is the difference between knowing something and believing it?" Since our senses are fallible—for example, our sight can be deceived by an optical illusion—how can we be sure about the things we see, touch, hear, smell, and taste? The great philosophers of history have faced these questions and attempted to answer them.

---

*At a much deeper level, there are certain differences between the two approaches that have never been fully analyzed. By exploring the meaning of these differences in much greater detail, it may be possible to gain new insights into the quantum theory.

But now the task had fallen to the physicists. Heisenberg's uncertainty principle was forcing the Copenhagen group to ask what it means to have knowledge and certainty about the atom. Indeed, they were beginning to ask whether atomic reality actually exists.

Classical physics offers complete certainty and predictability. Quantum theory suggests a reality in which no absolute knowledge is possible and in which indeterminism replaces determinism. Bohr felt that this new world had to be faced squarely and a consistent epistemology created. He began his investigation into quantum reality by thinking about how physics learns about the quantum world.

Since atoms are too small to be seen, how can we ever learn about them? The answer is to use laboratory apparatus that magnifies and registers tiny effects that occur at the atomic scale. For example, the click of a Geiger counter indicates that an elementary particle is passing by.

FIGURE 3–7

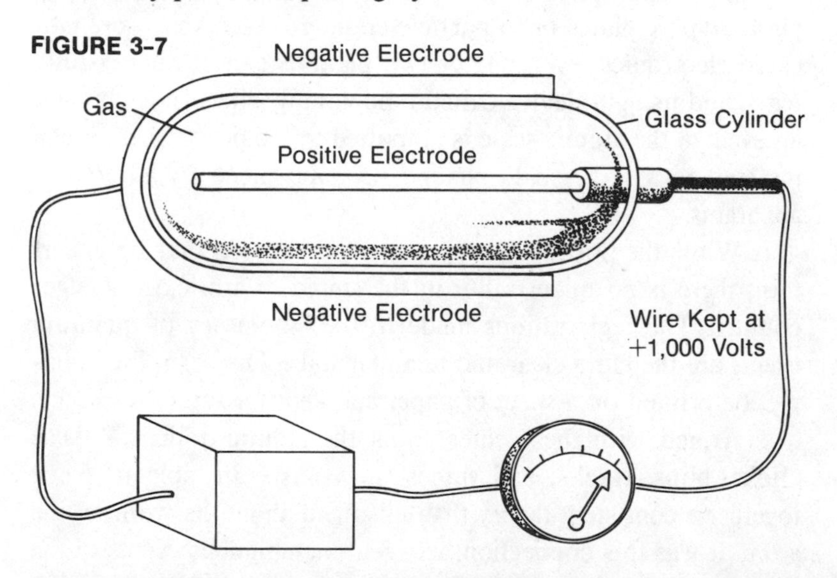

A Geiger counter. An elementary particle entering the gas strikes an atom and knocks out some of its electrons. The charged atom, accelerated by the electrical field in the counter, collides with other atoms. The result is an avalanche of charged atoms, which fly across the Geiger counter. This produces an electrical current, which is then amplified and displayed as an electrical pulse or audible click.

A Geiger counter is a hollow container out of which most of the air has been pumped. A wire in the center of the counter is kept at a very high voltage, almost to the point where a spark shoots across the counter. When an elementary particle rushes through the counter at high speed, it bumps into one of the atoms of air inside the container, stripping away some of its electrons and leaving the atom with an electrical charge. The charged atom is attracted by the charged electrical wire and accelerates toward it, hitting other atoms in the process. In turn, these atoms are stripped of some of their electrons, and they themselves begin to rush toward the wire. Soon the whole process becomes a cascade of collisions, with more and more atoms involved. The result is a tiny surge of electrical current through the gas, which is amplified by an electronic circuit to give an audible click. In essence, the Geiger counter operates on a hair trigger so that even a single elementary particle will cause it to fire.

Elementary particles can also be indicated by their tracks on photographic plates or in particle counters. Thanks to sophisticated electronics, even one or two photons can be successfully registered using a device called a photomultiplier. In each case, an event at the atomic scale is amplified to the point where it can register as a large-scale change in some piece of laboratory apparatus.

While the position or velocity of an electron may be uncertain, there is no uncertainty in the sound of a clicking Geiger counter. The registrations made in the laboratory of quantum events are therefore clear and unambiguous. They can, for example, be printed on a sheet of paper and kept for future reference.

Armed with these clues about the atomic domain—these clicks, blips, tracks, and traces—physicists are able to piece together a consistent theory that tells them about the world of the atom. It was this connection between unambiguous events in the human-scale laboratory and the ambiguous world of the atom that Bohr investigated. This was his point of entry into the epistemology of the atomic age.

Bohr pointed out that gathering all information about the world of the atom must, at some point, involve making a measurement with a laboratory instrument. To find out about an atom, or elementary particle, a scientist carefully sets up an experiment, with racks of electronic equipment, meters, dials, Geiger counters, and other detectors. The result of the experiment—which causes a dial to move or a Geiger counter to click—always involves an interaction between this large-scale apparatus and the quantum object. After all, if there were no interaction, if no transaction between the experimental apparatus and the atomic world took place, then nothing would have registered.

But every interaction must involve an exchange of energy; otherwise no change would be possible. And it is the basic premise of quantum theory that energy is always exchanged in indivisible units called quanta. No matter how refined the apparatus becomes, at least one quantum of energy must be involved. If no quanta are exchanged, then nothing has happened, and no change can be registered. But where even a single quantum is involved, the magnifying characteristics of the apparatus are geared to record a click or movement of a dial.

Everything we know about the quantum world therefore comes down to interactions between atoms, or elementary particles, and laboratory detectors and must therefore involve an exchange of energy. Energy performs work—it produces change—and if no net energy change occurs in an experiment, then nothing has happened, and nothing can be registered.

We already know that the quantum is indivisible, for this is the basic principle of quantum theory. Therefore, when an atom interacts with a detector, some interchange of energy is involved and this must involve at least one quantum or a series of one-quantum interactions. It is not possible to divide this quantum of energy in half and say, for example, that one-half of it comes from the atom and the other half from the detector, or that 90 percent comes from the atom and 10 percent from the detector. The quantum itself is indivisible and can never be broken into parts.

This means that we can never know exactly where any quantum comes from—we can never divide it into contributions made by atom and by apparatus—which implies that the moment the detector and the atom interact, the whole situation becomes an unanalyzable whole. It is unanalyzable because it is no longer possible to partition the situation into apparatus + atom—the two are bound together by a single, indivisible quantum. In Bohr's words, there is "an indivisible wholeness," an unanalyzable wholeness. At the moment of observation, the observer and observed make a single, unified whole.

This holistic nature of the atomic world was the key to Bohr's Copenhagen interpretation. It was something totally new to physics, although similar ideas had long been taught in the East. For more than two thousand years, Eastern philosophers had talked about the unity that lies between the observer and that which is observed. They had pointed to the illusion of breaking apart a thought from the mind that thinks the thought. Now a similar holism was entering physics.

But this holistic approach creates problems of its own. Whenever we attempt to discover the properties of an electron or proton, we must make a measurement. And whenever this act of measurement takes place, the apparatus and atom form a single, whole system. The only way we can learn about the atom is to enter into the world of undivided wholeness. As soon as we do this, however, we lose the whole notion of an independent atom or electron—it is bound to the measuring apparatus within an unanalyzable whole.

To illustrate the indissoluble link between observer and observed, Niels Bohr has given the example of the blind man and his cane. Where does the blind man end and the external world begin? For most of us, the world outside begins just outside the surface of our skin. But the blind man learns to feel through his cane, and for him the world begins at the cane's tip. Now imagine the blind man's sense of touch extending out of the tip of the cane and into the roadway itself. Imagine its extending along the road, to the passersby, to the trucks and cars, to the whole world. There would be no point at which the blind man ends and the world begins; all would be linked together. As with the blind man, so too with an atom or an electron: each time we attempt to observe it, we become linked to it so that we can no longer say which is us and which is the atom.

What, then, is an electron, a proton, or an atom? What properties does a particle have if it only manifests itself in an unanalyzable interaction with a piece of apparatus? What does it mean to say that the electron *has* a certain velocity or position if every attempt to measure these properties represents an irreducible act of interference? Indeed, it becomes a major problem to speak of the electron as "having" or possessing properties. And if all the properties of a quantum object become ambiguous, then what sort of a reality does it have?

Max Born's colleague Pascual Jordan declared that observations not only *disturb* what has to be measured, they *produce* it. In a measurement, "the electron is forced to a decision. We compel it *to assume a definite position*; previously it was, in general, neither here nor there, it had not yet made its decision for a definite position. . . . We ourselves produce the results of measurement."*

On the other hand, when we are not observing an atom or electron, quantum theory dictates that we cannot even talk about its moving along a given path, or even of its being anywhere at all! According to Heisenberg, while measurements and laboratory observations are real, "the atoms or the elementary particles are not as real; they form a world of potentialities or possibilities rather than one of things or facts."**

When a measurement is made, a quantum particle and the experimental apparatus are indissolubly united. When the measurement has ended, the quantum particle is on its own. The

*Quoted by Max Jammer in *The Philosophy of Quantum Mechanics* (New York: John Wiley, 1974).

**W. Heisenberg, *Physics and Philosophy* (New York: Harper Torchbooks, 1962).

clicks of the Geiger counter enable us to say, for example, that an electron just passed through a particular region of space. A moment later, a second Geiger counter may register that electron's arrival.

But where was it in the short period between? Somewhere between the two? Somewhere in the laboratory? Somewhere on earth? Somewhere in the solar system? In fact, quantum theory does not permit us to answer any of these questions. Since the theory is indeterministic, there is no sense in talking about the electron as "having a path." Just because an electron was at point *A* one moment and at *B* the next, we are not justified in assuming that it actually took some path to move from *A* to *B*! All that is real, Niels Bohr and his colleagues would say, are the two Geiger counter measurements. What happened in between is pure mystery.

This, in essence, is the interpretation of quantum theory proposed by Bohr and his colleagues. But in creating a consistent account of the atomic world, they almost seem to have thrown the baby out with the bathwater. By refusing to talk about atoms and electrons when they are not being observed, by refusing to entertain such notions as paths and trajectories, they seem to have left the theory almost empty of content.

Pauli talked about what he called "the irrational in nature." Quantum theory was forcing scientists to accept that irrational element. Bohr for his part argued that every time we attempt to go beyond this quantum austerity and build pictures of the atomic world, or to create models in the mind, then the ideas of classical physics will creep in. All talk about paths, orbits, and intrinsic properties represents hangovers of classical thinking and traditional ways of visualizing the universe. As soon as we try to imagine pictures of the atom, such ideas enter, and the result is paradox and confusion. The best we can do is to cut ourselves free from it all.

Then where is atomic reality? Heisenberg suggested that the reality now lies in the mathematics. The formulas of matrix mechanics or wave mechanics work perfectly well. They can be used to account for any measurement in the laboratory. If you want to know where atomic reality lies, then Heisenberg points to the equations; there is no hope of finding it anywhere else.

Does this mean that there is no reality outside the mathematics, or simply that mathematics is the only way in which we can consistently discuss quantum reality? About this, Bohr is at his most uncompromising. "There is no quantum world," he says. "There is only an abstract quantum mechanical description."* While the French physicist and philosopher Bernard d' Espagnat had referred to the quantum world as a "veiled reality," Bohr went even further: there *is* no quantum reality, he said. "Don't look at the man behind the curtain," cried the Wizard of Oz as Toto discovered that all the wizard's illusions of sight and sound were nothing more than the production of a carny's box of tricks. Reality in the land of Oz was purely the creation of the wizard, but in the quantum world, Bohr argued, there is no wizard. There is nothing behind the curtain; all we see are phenomena, the tricks themselves.

Bohr argued that if we must ask for descriptions, then the best we can settle for is to accept the approach of complementarity. There is no single, unambiguous account for the quantum world. If we want to hang on to words like *wave* and *particle* or *space* and *time*, which are all classical concepts, then we also have to accept that these words must now be used in complementary and paradoxical senses. If an electron is a particle, it is also a wave. It is at once discrete and distributed, localized and wavelike.

If we still want to think in terms of pictures as we enter the atomic world, Bohr said, then we must face paradox as well. Once we have accepted the implications of quantum theory, there can be no turning back. The classical world has been left behind forever.

---

*M. Jammer quoting an account of Bohr in *The Philosophy of Quantum Mechanics* (New York: John Wiley, 1974).

4
BOHR vs. EINSTEIN

Quantum theory has changed the way scientists think about the world. It has given a new meaning to the idea of an "objective reality" at the subatomic level. Working together in Copenhagen, Bohr, Heisenberg, and Pauli had constructed what they believed to be a complete and entirely consistent account of the atomic world. It was their intention that the Copenhagen interpretation should put an end to debate about the meaning of the quantum theory and avoid the paradoxes that their opponents had proposed.

But not everyone accepted what Bohr and his colleagues had achieved. Several major scientific thinkers of the day refused to believe that quantum theory was the final word on atomic reality. One of the most outspoken criticisms came from Albert Einstein, who argued that objective reality must extend right down to the quantum level. Einstein was also unhappy with the indeterminism of quantum theory in which quantum events have probabilistic origins with no deterministic underpinning. But it was the lack of a quantum reality that most concerned Einstein. Surely, he argued, there must be a deeper theory, as yet undiscovered, that will reveal the independent reality of the atom.

In the first years of this century, Einstein had been responsible for one of the most revolutionary changes in physics—the theory of relativity. He had introduced a world in which space and time are unified, in which moving clocks run slow, and moving objects appear shorter. He had unfolded the famous twin paradox, in which twins going on different journeys age at different rates. In short, Einstein had shown that if we travel close to the speed of light, the whole appearance of the universe changes. Why should a scientist who believed in the relativity of appearances be so worried about retaining an objective reality?

Einstein firmly held that beyond the world of appearances there exists an independent, objective reality. Whatever scientists measure and observe may be a function of the objects' velocity and location in the universe, yet these appearances are nevertheless founded on an underlying objective, independent reality. It may never be possible to obtain a direct perception of that reality—for appearances are relative to the observer—yet it is possible to grasp some essence of that reality through science.

Reality, according to Einstein, manifests itself in physical laws. The laws of nature are expressions of patterns in the reality of the world, and these laws are totally objective and independent of the particular state of any observer. Although experiments and observations yield relative appearances, it is possible, by constructing theories, to reach beyond appearance and touch the deeper laws of reality. Newton's physics had been one attempt to reach the laws of reality, but it had been flawed. Einstein's approach was deeper; it brought us closer to the truth, closer to the independent, objective reality of the world.

While the theory of relativity decrees that different observers see different events, nevertheless it brings us closer to the underlying objective laws of nature. These laws, Einstein said, are deterministic and entirely unaffected by the actions or particular state of being of any observer. We may change the appearances of things by moving at different velocities, but we cannot change the laws that govern these appearances.

This reality of Einstein's is known as a "local" reality. Each system or object can be defined and understood in its own particular region of space. Objects have their own independent existence, and if they change, then it must be as the result of interactions or forces acting from outside. These forces can also be defined in an objective way through the laws of nature. There are no mysterious "actions at a distance," no mystical influences. Objects possess properties, Einstein said. They move along paths; their fates are determined.

But then Bohr and Heisenberg came along with complementarity and the uncertainty principle. They denied that the electron has a path, or even that it possesses any intrinsic properties like position and velocity. In fact, they seemed to be saying that the only reality one can talk about lies in the mathematical equations, and that there is no point in trying to construct mental models of the quantum world. To Einstein this was a "tranquilizing philosophy" (as he put it in a letter to Schrödinger), a metaphysical approach to the world that induced a sleep of the mind by smothering questions about the ultimate nature of the quantum world. The Copenhagen interpretation was nothing more than "a soft pillow on which to lay one's head"; it was not a true theory of nature or an attempt to engage reality face to face, but an encouragement to daydream.

Einstein and Bohr met, on a number of occasions, at the famous Solvay Conferences in Brussels, where a small group of the world's leading scientists would congregate each year. During these meetings, Einstein pressed his objections to the new quantum theory. Time after time he would propose ways in which an objective meaning could be given to events and processes in the quantum world. It was unthinkable that the idea "reality" should come to an end at the scale of an atom. Bohr, however, stressed that *reality* is merely a word in our language. Almost echoing the great philosopher Ludwig Wittgenstein, although Bohr apparently arrived at his insights independently of the Austrian-born philosopher, Bohr pointed out that the trick was to learn to use the word correctly. (For Wittgenstein, many of the "great philosophical problems" were really nonproblems, confusions generated by misuse of language and laziness in distinguishing one word game from another.)

Einstein was an intuitive yet down-to-earth thinker. He therefore phrased his arguments in terms of hypothetical experiments designed to point out holes in Bohr's careful interpretation. For example, Heisenberg's uncertainty principle had set a limit on what could be known about a quantum object. Einstein was quite willing to accept that nature may put practical limits to what we can know and measure about the world. But this did not mean that a deeper, objective reality itself does not exist.

Bohr, in contrast, had gone further by arguing that even the concept of the electron having a simultaneous position and velocity has no meaning in the theory. Choose one of two complementary descriptions—velocity, say—and you sacrifice the other. Einstein, by contrast, attempted to give meaning to these so-called ambiguous ideas by devising ingenious "thought experiments" in which it should be possible to tease out a little more information from nature.

Einstein's thought experiments were presented to Bohr, who, after a few days' thought, was always able to discover some flaw in the argument, some way in which nature would conspire to keep her secrets hidden. No matter how ingenious Einstein became, his proposal would always contain a subtle flaw.

The thought experiments and arguments that Einstein presented are subtle and carefully thought out. While it would take too much time to go into them here, the illustrations below indicate the concrete level at which Einstein and Bohr were thinking. In this case, the three pieces of apparatus illustrate an argument about the double-slit experiment. Einstein was attempting to reduce the quantum uncertainties involved by trying to take into account the precise interactions with the experimental apparatus. In each case, Bohr was able to show that some other uncertainty would creep in to frustrate Einstein's argument.

FIGURE 4-1

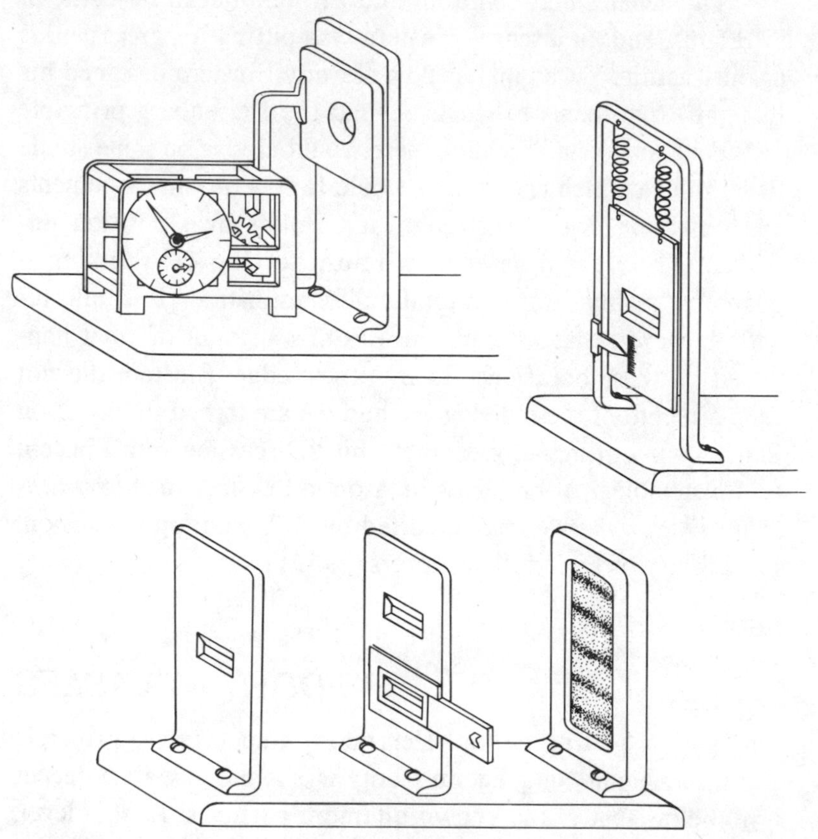

In his discussions with Niels Bohr, Einstein proposed a variety of thought experiments aimed at getting around the restrictions of quantum uncertainty. In turn, Bohr would sketch the apparatus in order to focus the discussion. (a) Each piece of the apparatus used in the double-slit experiment is firmly bolted down so that there can be no uncertainty about the position of the particle. But it is then no longer possible to control or determine any changes in momentum (mass × velocity) as the particle passes through the apparatus. (b) By designing a slit in this way, it becomes possible to measure exchanges of momentum (mass × velocity) as the particle passes through the slit. But a movable slit introduces uncertainty into the particle's position. (c) This piece of apparatus, used in a different experiment, was designed to pin down the exact time a particle passes through a tiny hole.

The debates between Bohr and Einstein became a battle of the giants, and, in a sense, Einstein was pitting his great genius against nature. Yet no matter how cleverly Einstein designed his thought experiments to reach beyond the Heisenberg principle and touch an essential reality, there would always be some subtle flaw through which uncertainty would flow. Einstein's arguments and objections were frustrated, and Bohr's own position unshaken. There could never be a return to classical thinking—quantum "reality" was of a totally different nature. [It is unfortunate, however, that we have only Bohr's account of what happened on these occasions. To my knowledge, Einstein did not keep a record of these dialogues, and we are forced to rely upon Bohr's own writings—for example, his "Discussion with Einstein on Epistemological Problems in Atomic Physics" in *Albert Einstein: Philosopher-Scientist*, edited by P. A. Schilpp (Evanston, Ill.: Library of Living Philosophers, 1949).]

HIDDEN VARIABLES

One way of getting beyond Heisenberg's uncertainty principle and the ambiguity of quantum reality was to propose a yet deeper level below that of the known quantum particles. At this level, perhaps built on even smaller particles, the idea of an objective reality would be restored.

Something analogous to this had already happened in physics. In the eighteenth century, chemists and physicists studied the properties of various gases as their pressures and temperatures were varied. A series of laws, called the gas laws, were discovered, showing for example, how the volume of a gas increases with temperature. Yet no one knew the deeper reason for these particular laws or what lay behind them.

It was only when scientists proposed that everything is made out of atoms and molecules that the gas laws made sense. Simple calculations showed that if gas molecules behaved like tiny billiard balls, speeding and colliding with each other, then they would produce exactly the same relationship between pressure,

FIGURE 4-2

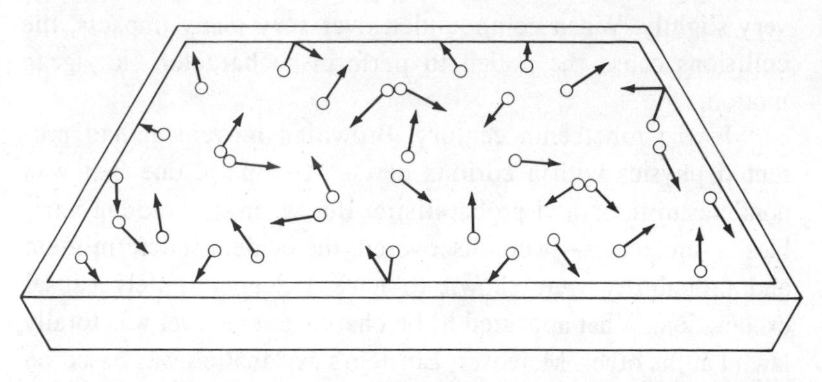

Gas molecules collide with each other and with the wall of their container. The effect of the myriad collisions with the wall is to produce pressure. The higher the temperature, the faster the molecules are moving, and the greater the pressure.

temperature, and volume as was expressed in the gas laws. Molecules could be thought of as the previously "hidden variables" that made sense out of the observed properties of gases.

Could there be some new hidden variables that would likewise make sense out of quantum jumps, indeterminism, and the paradoxical double-slit experiment? Was the ambiguous quantum reality based on some even deeper subquantum objectivity?

In fact, Einstein had solved a related problem for his Ph.D. thesis—determining the hidden variables that govern what is called Brownian movement. In 1827 Scottish biologist Robert Brown noticed, through his microscope, a continuous zigzagging movement of tiny pollen grains in water. At first he believed that this dance had something to do with reproduction, but later he observed it in plants that had been dead for over a century and even in tiny particles of smoke and chips of glass. This continuous Brownian movement appeared to be some sort of perpetual motion. What could be its cause?

In the end it was Einstein, and independently Marian Smoluchowski, who discovered the hidden cause—molecules. Water molecules are so tiny as to be beyond the range of any optical microscope. In addition, by virtue of the finite temperature of the water, they are in constant motion, making tiny impacts with the pollen grains. Each molecular impact causes the grain to shift very slightly. When compounded over very many impacts, the collisions cause the pollen to perform a characteristic zigzag motion.

In the nineteenth century, Brownian movement had presented physics with a curious new sort of path, one that was nondeterministic and probabilistic. But when its "hidden varibles"—molecules—were discovered, the evident indeterminism and probability were shown to have a deeper, purely causal explanation. What appeared to be chaotic at one level was totally lawful at the other. Moreover, Einstein's explanation was based on the idea of a "local reality"—the overall path of the pollen grain being determined by the local details of what happens at each point in space and time.

Does something similar exist within the quantum world? Are there underlying sub-elementary particles and forces, invisible to elementary-particle accelerators but nevertheless determining events at the level of atoms and elementary particles? If this were true, then determinism and an independent local reality could be restored to physics.

Suppose entirely new sorts of particles exist that obey deterministic laws. The complicated movements of these particles' interactions may give rise to phenomena that we interpret as electrons along with their wave-particle duality, quantum jumps, and double-slit behavior. At one level, things may appear probabilistic—yet at another, deeper level, they would be deterministic. Furthermore, these variables would be so extremely sensitive that any attempt to measure one combination of them would cause unpredictable interactions that would totally change the result of other measurements. In this way, Heisenberg's uncertainty principle would originate in the very sensitive arrangement of different configurations of hidden variables.

Ideas like these may from time to time have occurred to Einstein; they were certainly proposed by other scientists who could not accept quantum theory. If hidden variables—new sorts of hidden particles and interactions—truly existed, then this would mean that quantum theory was no longer a complete theory of the subatomic world. It would be simply a stopgap hypothesis on the way to some deeper theory. This hidden-variable theory would be deterministic and allow for an objective, independent reality. In such a theory, a quantum object has a definite velocity and position at every instant of time. These properties are "possessed" by the quantum system and entirely independent of any observer.

Hidden variables are one way out of the quantum dilemma. But do they actually exist in the quantum world? The final verdict had to await the coming of Bell's theorem. And Bell's theorem originates in an earlier, and remarkable, paradox proposed by Einstein, Boris Podolsky, and Nathan Rosen.

THE EINSTEIN, PODOLSKY, AND ROSEN PARADOX

According to the Heisenberg uncertainty principle, any attempt to measure both the position and the velocity (momentum) of an electron must lead to uncertainty. Einstein was perfectly willing to accept the conclusion that quantum events are so sensitive that an attempt to measure one property will unpredictably change the other.

But Einstein could not accept the Copenhagen interpretation's denial of any meaning to the very idea that a quantum particle possesses a definite position and velocity. Instead, he maintained that reality has an objective meaning, so that an electron can simultaneously possess a definite position and velocity, even if these properties can never be measured exactly.

Einstein did not pin his hopes on hidden variables or any other specific alternative—rather, he simply objected to the claim that quantum theory was the final and uncompromising word on subatomic reality. The example of hidden variables does, however, provide a graphic way of understanding how a deterministic reality could produce unpredictable and complementary results in a series of measurements.

In Einstein's universe, it is as if everything can be characterized by cards of identity. Like international travelers, quantum systems carry identity cards that specify all their properties. Nature may conspire to prevent our reading all the information on these cards, but it nevertheless makes sense to speak of the independent reality of a quantum system. It was this notion of an independent reality composed of a variety of objective elements—cards of identity—that Bohr denied. For him the quantum world had no objective reality apart from our acts of measurement, and different experiments forced it to give different, complementary answers.

Einstein rejected this view. "Does the moon exist only when you look at it?" he once asked his colleague Abraham Pais. Since he could not accept Bohr's position, Einstein found it necessary to create yet another thought experiment that would demonstrate that objective reality has a definite meaning. This time his argument would be more difficult to refute. His ideas were crystallized in discussions with two colleagues, Boris Podolsky and Nathan Rosen, and published in the paper "Can Quantum Mechanical Description of Physical Reality Be Considered Complete?" which appeared in the American physics journal *Physical Review* in 1935.

In the words of Léon Rosenfeld, the paper fell upon the Copenhagen group "like a bolt from the blue." Bohr and Rosenfeld spent weeks grappling with its central arguments as they attempted to discover a logical flaw. If Bohr and his Copenhagen interpretation were correct, then Einstein, Podolsky, and Rosen just had to be wrong.

What the three physicists (let's call them EPR for short) had done was to describe a hypothetical experiment in which it makes perfect sense to talk about the simultaneous velocity and position of quantum particles—in other words, a state of affairs in which position and velocity have definite and precise meanings that are quite independent of any observer. Quantum theory, of course, refused to address such "independent elements of reality," as EPR called them; indeed, the Copenhagen interpretation claimed that such ideas were ambiguous. EPR therefore concluded that quantum theory must be incomplete, since it could not account for parts of reality that have a clear existence independent of any observer. Quantum mechanics, they deduced, was a passing theory, a stopgap on the way to something deeper.

But how could EPR argue that independent elements of reality, like position and velocity, have a precise and independent existence? After all, any attempt to measure these properties is always accompanied by an irreducible disturbance. The EPR argument can be illustrated by means of a fairy story. Twin brothers, sons of an important king, have been brought up together in a secret castle. These two princes, who talk and dress alike, ride out on a great adventure. Prince Nathan travels west, while Prince Boris rides east. After traveling for a day and a night, Prince Boris is captured by the wizard Albert, who imprisons him in a tower. The wizard desires to discover all he can about the second brother, Prince Nathan, so that he can ensnare him. He reasons that since Nathan and Boris are identical twins and have been brought up in the same way, he can find out about Nathan by questioning Boris.

Prince Boris is put to the rack and tortured. At first he resists with great fortitude, but later he begins to talk. There is a painful disturbance of the system (Boris) by the means of observation (the rack), but, the wicked wizard argues, this disturbance can have no effect on Nathan, who is two days' journey away and can know nothing about the interrogation. The answers that are extracted from Boris, destroying him in the process, allow the wizard to build up a picture of Nathan. The wizard has gained knowledge about Nathan's "reality" without ever disturbing his existence. The wizard Albert can now ride out on the trail and enchant Prince Nathan.

The thought experiment planned by EPR works in a similar way. Since a measurement disturbs a quantum system, the idea was to perform the measurement on one quantum system and use its result to deduce information about its twin some distance away. In the EPR experiment, two similar quantum systems, *A* and *B*, are allowed to interact until their internal states become correlated. They then separate until they are at opposite sides of the laboratory. Performing experiments on *B*, EPR argued, will make it possible to deduce information about *A* and in such a way that *A* is never disturbed.

Suppose that two electrons are used. They start off at rest in the center of the laboratory, then move away from each other. For example, the two electrons may be shot off in opposite directions as the result of some nuclear disintegration. An important law of nature, which holds for both classical and quantum systems, declares that if the two electrons begin at rest, then the velocity of one will always be opposite to the velocity of the other. The actual value of these velocities is unknown, but scientists can be assured that their sum will always be zero. In other words, once one velocity is measured, the other can be deduced.

Suppose we measure the velocity of *B*. As soon as we have done this, we can deduce the velocity of *A*. In EPR's terminology, this velocity is now an unambiguous element of the reality of *A*. But since we have done nothing to *A*, have affected it in no way, this must mean that *A* must always have been in such a state, with this certain velocity. This particular element of reality has always been registered on one of *A*'s cards of identity, and now we happen to know its value.

But suppose we had decided to measure the position of *B*. Again it becomes possible to know *A*'s position without affecting it in any way. But this implies that the position of *B* is an independent element in *B*'s reality and something that has not been produced by any measurement. Both position and velocity are "cards of identity" that have a very definite physical existence. Admittedly it will never be possible to measure either of these properties directly without disturbing the other. But it does imply that such properties can have a definite, objective meaning. But precise knowledge of these two independent elements of reality violates Heisenberg's uncertainty principle. Quantum theory cannot account for this; the theory is therefore incomplete, EPR concluded.

EPR were not so much concerned with any practical attempt to know directly about *A*'s reality. After all, any attempt to verify their prediction of one of *A*'s "cards of identity" would involve making measurements on *A*, and these very measurements would create irreducible disturbances of *A*'s reality. The whole key to the EPR experiment is to leave *A* alone, to avoid interacting with it in any way yet, and by means of measurements on its correlated partner *B*, to make deductions about *A*'s independent reality. EPR believed that they had thus given a definite meaning to talking about the position and the velocity of *A*, something Bohr and his Copenhagen interpretation had always denied.

This was the bombshell that Einstein, Podolsky, and Rosen burst in the Copenhagen camp. "They do it 'smartly,' " Bohr agreed, "but what counts is to do it right."

Bohr refuted EPR in a way that turned out to be quite subtle. It was not so much that there was any error or flaw of logic in the EPR argument. Rather, the argument was not strong enough or relevant enough to convince the Copenhagen school. To put it another way, the assumptions and deductions of Einstein's argument were insufficient to overturn the Copenhagen interpretation. In arguing their thought experiment, EPR were making the same assumptions about the nature of reality that Heisenberg, Bohr, and Pauli thought they had left behind them a decade before.

Bohr's refutation of the EPR argument begins by emphasizing yet again that quantum mechanics demands a radical change in what we mean by physical reality. In fact, he claims, the whole notion of separate "elements of reality," as applied to quantum systems, no longer has a precise meaning. What *is* important is the whole set of conditions under which a quantum experiment is made. Choose one set of conditions, and a particular aspect of the quantum system is revealed. Choose different conditions, and a complementary aspect is exhibited.

Take, for example, the velocity of an electron. In Bohr's opinion, this is not some objective property that the quantum particle possesses like a card of identity. Rather, it refers to a particular experimental arrangement designed to reveal an aspect of the electron. With such a setup, it makes sense to talk about the movement of a dial as signifying the velocity of the electron. But we should not immediately assume that this velocity has all the familiar properties associated with classically moving objects. The meaning that must be attributed to the velocity of the electron at *B* in the EPR paradox lies in its relationship to a particular experimental arrangement in the neighborhood of *B*.

Once the velocity of *B* is measured, it is admittedly possible to predict accurately what would be registered if a similar measurement were made at *A*. But this does not mean than *A* "possesses" that velocity—that a particular velocity now forms an independent element of *A*'s reality. Only when an actual measurement has been carried out at *A* does it make sense to talk about *A*'s velocity.

Position is another property that can be measured at *B*. It is complementary to velocity, which means that the one experimental arrangement is incompatible with the other. The decision to measure position rules out any hope of determining velocity, and vice versa. To be specific, once the *decision* has been taken to measure the position of the system *B*, then the configuration of that particular experimental apparatus precludes knowing anything about the electron's velocity (i.e., setting up one experimental situation is incompatible with the other). Moreover, there is no point in talking about the velocity of *B* under such conditions, since "velocity" is not something an electron possesses in any objective way.

In the second part of the EPR experiment, an entirely different experimental arrangement is required to measure position. With this new knowledge, of the position of *B*, it becomes possible to predict the result of a measurement for position that could be made at *A*. But this does not mean that a particular position has become a property of *A*. *A* has no real position, Bohr would argue; the concept is entirely inappropriate to an isolated quantum particle when it is not interacting with a piece of experimental apparatus. Until an actual measurement is performed on *A*, it makes no sense to speak of *A*'s position.

Indeed, in Bohr's opinion, it has never made sense to speak, as EPR do, of the simultaneous velocity and position of *A*, or of *A* as possessing independent elements of reality. Such concepts have no meaning in quantum theory. Bohr's refutation of the EPR paradox has nothing to do with disturbances or mysterious influences passing between one system and the other. Rather, it rejects the very assumptions of independent reality made by Einstein and his colleagues, the assumptions on which their whole thought experiment is based. Bohr had no objection to predicting the results of measurements that may be carried out at *A*, based on knowledge gained at *B*. But this is very different from saying that the electron possesses both a well-defined position and velocity.

Once we renounce the classical idea of an independent reality in which systems have well-defined properties, Bohr would say, then the paradox proposed by EPR vanishes.

This, in essence, is Bohr's refutation of Einstein's position. The problem was that it did not satisfy Einstein. In fact, Bohr's argument does not really meet Einstein's objections any more than Bohr thought the EPR paradox undermined the Copenhagen interpretation. Both parties have failed to engage the other or to address what their opponent felt to be the essential feature of their argument.

The key, as Bohr stated, was "essentially the question of an influence on the very conditions which define the possible types of predictions regarding the future behavior of the system." Bohr appears to have been talking about the particular experimental arrangement that would allow us to interpret a measurement as referring to, say, velocity and would thereby rule out some other complementary, noncommuting property.

Einstein, however, must have puzzled over this phrase, as indeed have the generations of physicists who came after him. What is Bohr talking about, Einstein must have wondered, what does he mean? In view of the great distance between *A* and *B*, "an influence on the very conditions" has nothing to do with mechanical forces or other sorts of pushes and pulls that act on a distant object. Bohr provided a reply that cannot have meant much to Einstein.

The situation between Bohr and Einstein is not unlike what happens in a long-suffering marriage in which each partner complains that the other never listens. The husband attempts to voice his grievances to a nagging wife. The wife is ignored and her needs neglected by the husband. Both parties talk a great deal yet are frustrated by never making contact at a deeper level. Each fails to engage the other's concerns, and their dialogues involve endless deflections and confusions. Something similar was happening in the heart of physics.

A more revealing metaphor is a hypothetical conversation between Cézanne and a realistic painter of the French Academy. Both have been working on a still life—a cloth-covered table with an arrangement of oranges. "But," the realist complains to Cézanne, "your picture is unrealistic. It lacks truth. I have sat before that table day after day and studied the way the light falls on the oranges. I know every fold of the cloth, my draftsmanship is perfect, my color sense is excellent."

"I too have looked at that table till my eyes ache," Cézanne complains. "I have walked around it, picked up an orange, and stared at it from every angle. I have grasped the truth of every fold of the cloth. Don't accuse me of being untruthful."

Each feels that the other has missed the essential truth of the scene, yet there is little chance that they will ever agree on the nature of artistic truth. The realist can point to the fact that Cézanne's table appears to break in the middle and lacks continuity, that an orange close to the viewer is not painted to appear smaller than one that is farther away. Cézanne, for his part, can complain that the "realistic" painting presents only a single frozen viewpoint and never really engages the solidity of the scene or the fact of its existence in three-dimensional space.

The two painters have different views of what constitutes truth in painting, and this arises from their different passions. Each painter is visually engaged with nature in very different ways and each gives great value to certain kinds of exploration while being relatively indifferent to others. As a result, they produce quite different versions of "reality." Niels Bohr would have understood this predicament very well. He would have called it complementarity.

But no such complementarity was permitted when it came to the acceptance of quantum theory. For Bohr there could be only one possible interpretation, and Einstein was incapable of accepting it. Bohr never really addressed Einstein's difficulties in accepting the quantum theory. Likewise, Einstein could never release his grasp upon the notion of an independent reality long enough to communicate with Bohr.

Einstein was willing to make compromises and go beyond his original position. He wondered, for example, whether the traditional concepts that had been used to define classical systems, such as velocity and position, no longer apply at the quantum scale of things. Perhaps entirely new concepts were needed; perhaps new physical laws govern the behavior of quantum systems. It may even be that Heisenberg's uncertainty principle prevents any exact knowledge of these quantum concepts. Nevertheless, Einstein believed that these unknown states and concepts have a definite physical reality. Proposals like this, by Einstein, could never meet the challenge set by Bohr.

As Pauli put it, "One should no more rack one's brains about the problem of whether something one cannot know anything about exists all the same, than about the ancient question of how many angels are able to sit on the point of a needle. But it seems to me that Einstein's questions are ultimately always of this kind."* Einstein continued to hang on to reality at all costs while Bohr pointed out that *reality* is just a word in our language and therefore must be used correctly.

Despite their insights, Einstein and Bohr were becoming increasingly unable to play their word games together. So the two great men battled on, each failing to engage the other. It was

---

* Pauli in *The Bohr-Einstein Letters* (New York: Walker, 1971).

inevitable that what had begun as an intense personal and intellectual friendship should eventually peter out. After all, there was no real point in Bohr and Einstein talking to each other anymore.

Physicist David Bohm has often spoken of the party given at Princeton in an effort to bring Bohr and Einstein together. Both scientists were invited, yet Bohr and his colleagues stood at one end of the room, with Einstein and his colleagues at the other. What is most tragic is that this personal drama between Bohr and Einstein has also been played out within physics itself. Deep within the heart of that discipline, there remains an area of confusion and conflict.

5
BELL'S THEOREM

With the EPR paradox, Einstein had taken his best shot at the quantum theory, and in Bohr's opinion, that shot had been deflected. Quantum theory had emerged intact, and the Copenhagen interpretation had, if anything, been strengthened.

Here the story should have ended, with quantum theory accepted just as relativity had been a few decades earlier. The problem was that scientists continued to argue about what it really meant. The old problem of quantum reality had not been put to rest. Although Bohr believed that the Copenhagen interpretation resolved all outstanding difficulties about the meaning of the quantum world, physicists kept asking questions.

Bohr had said, "There is no quantum world. There is only an abstract quantum mechanical description." Heisenberg had echoed, "The atoms or the elementary particles are not real; they form a world of potentialities and possibilities rather than one of things or facts." In their view, quantum systems do not have an independent reality, they cannot be said to carry cards of identity or to possess properties when they are not being observed. Indeed, Jordan said, "We ourselves produce the results of measurement."

But what about the meters, dials, equipment, and even the laboratories that scientists use in their everyday work? No one would disagree about their reality, yet they are ultimately made out of molecules, atoms, electrons, photons, and so on. Laboratory apparatus, and indeed entire laboratories, are composed of quantum systems that in themselves are no more than ambiguities and potentialities. Yet these aggregates and compositions of ephemera are what are supposed to give reality to the quantum world every time a measurement is made!

Is there a paradox in this? Does reality begin only at some particular scale of size? Or is it a matter of complexity? Does the particular way in which the multiplicity of quantum systems that make up a stopwatch interact make time real? Or could it be that quantum mechanics applies only on the small scale and that its laws break down at our everyday level of things? If reality dissolves away at the atomic level, then on what foundation does our everyday world rest?

In fact, this problem extends to the whole universe. Physicists now believe that the universe began in a big bang followed by an incredibly short period of explosive expansion. But what was the origin of the big bang itself? Current thinking is that everything we know owes its existence to a fluctuation in some primordial wave function. It was one of these fluctuations that produced the initial conditions of our universe.

Yet the wave function is not real; it is simply a device used in the mathematics of quantum theory. Indeed, it is a wave of probability rather than an oscillation of matter. What the wave function describes is the probability that a particle will be discovered in a particular region of space should a measurement be carried out. This same wave function is also the mathematical tool used to predict the outcome of other experimental measurements. So what sense does it make to talk about the reality of wave functions and quantum states when no laboratory apparatus is around—indeed, when no large-scale world yet exists?

Physicists want to describe the creation of the universe in

quantum mechanical terms, and this means using a wave function. But what can be said about the initial instants of a universe when only elementary particles and photons of energy are present? In the absence of laboratories and observing scientists, what reality did the newborn universe possess?

While the Copenhagen interpretation remains the official "party line" of physics, many physicists feel that this abstract account does not really answer their questions. As Basil Hiley, the longtime collaborator of David Bohm, puts it, "Scientists come to praise Bohr and decry Einstein. But they end up thinking like Einstein and ignoring Bohr."

For two hundred years, physicists have been attempting to discover the nature of reality. It is hard to accept that, at some fundamental level, this reality ceases to exist. What is there left to grasp? Where is the certainty in a world of probabilities and potentialities?

Some scientists suggest that a "cut" or discontinuity exists between the quantum world and our large-scale reality. In other words, two descriptions of nature are needed, one for the large scale and the other for the atomic. Quantum theory serves for a discussion of elementary particles and atoms. But when it comes to the large scale, we need the world of Newton and Einstein. The only problem is where to place that cut between classical reality and quantum ambiguity. Is it a hard-and-fast cut, or does it move depending upon the nature of the system? And how is the nature of this cut decided in reality? Does it emerge out of the quantum theory, or are new ideas and concepts needed?

Jordan said, "We ourselves produce the results of measurements." But what exactly is the nature of this "we"? Does it include the human observer, or is a piece of laboratory apparatus sufficient to generate a measurement?

In the conventional Copenhagen interpretation, it is the decision taken by the human scientist that determines which of two noncommuting observables is measured. There is nothing mystical about this; it is no different from saying that the decision as to which aircraft to board determines your destination. While a human decision is subjective in the sense that it determines what aspect of a quantum system will be realized, it has nothing to do with the physics of the measurement itself. The velocity or the position of an electron can equally well be recorded on a printout or in a computer memory. Provided that this value has been registered on some piece of large-scale laboratory apparatus, then its existence is totally objective and does not depend on any human observer. Indeed, long before humans existed on earth, events have had definite outcomes. Quantum processes occur in stars; nuclei decay in radioactive uranium. Provided that these various quantum events are "registered," by producing some sort of change in the environment, they become totally objective and real.

Some scientists, however, argue that human consciousness does play a special role in determining quantum reality. They believe that a certain quantum fuzziness infects all laboratory apparatus, so that definite, objective outcomes are possible only with the intervention of a human observer. Hook a measuring device to a radioactive atom, and it too resides in an overlapping state in which both possibilities—disintegration or no disintegration—coexist. Or, rather, the potentialities for two sorts of different results still exist; the apparatus itself does not actualize either.

Consciousness or human mind is not, however, touched by this quantum uncertainty and is able to transform potentialities into actualities. The human observer of the quantum measurement is able to collapse the apparatus into a definite state and thereby cause it to register an objective result. But what if a piece of apparatus is attached to a printer? Some physicists would argue that the actual printout exists in an ambiguous state until it is observed. Others go as far as to suggest that the cortex of the human observer who views this readout is similarly suspended between several physical states.

But these are not the only alternatives to the Copenhagen interpretation. Some physicists reject the significance of human consciousness and ambiguous realizations in favor of a totally realistic interpretation—but in a variety of parallel universes. Advocates of this interpretation reject veiled realities, uncertain outcomes, and ghostly realities in favor of hard-and-fast objective realities. When a quantum measurement has several potential outcomes, this interpretation proposes that each outcome is actually realized, but in a different universe. Each quantum measurement is a sort of bifurcation point between universes. At the moment of measurement, our universe fragments into a number of almost identical copies. Each different universe contains all the memories and structures of its common origins and, in addition, one of those particular quantum outcomes. A little further down the line, each of these universes will split again, as yet other measurements are made.

Reality therefore is multifaceted; it consists of a potentially infinite number of parallel universes. Each universe is totally real to its inhabitants, who are incapable of interacting with or ever knowing about the universes that coexist beside them. Close to their bifurcation point, these universes will be remarkably similar, but as time goes on, some of them may diverge as a result of a series of quantum decisions.

Back at the big bang origin of the universe, at the moment (or nonmoment) of fluctuation of the primordial wave function, many universes may have been created. These continue in time, constantly bifurcating and splitting off. Many of them contain stars, galaxies, worlds, plants, animals, and consciousness. A vast number of them contain earths, humans. Indeed, many of these universes contain worlds in which copies of this book are printed. Some may not. (Royalty payments, however, are at present confined to a single universe.)

But with all these bifurcations of our world going on, why don't we feel something? The answer is that each individual consciousness bifurcates along with the universe. When an indeterministic quantum system encounters a measuring apparatus, it is not so much forced into a decision as the whole system of quantum + apparatus realizes each decision. And, since a number of different outcomes are possible, a number of different apparatus now exist, each registering a particular outcome. But that apparatus is coupled, in complex ways, with the universe, and each individual consciousness is part of that universe. So, at the moment of actualization, consciousness splits, and a number of parallel minds are born, each in its own universe. Subjectively, nothing much has happened. Each individual experiences a total continuity of experience and memory, not realizing that he or she has passed into a multiple mirror, with each reflection in a separate, self-contained universe.

The idea of a potential infinity of alternative universes sounds like pure science fiction, yet it has been adopted by many physicists. It does, after all, solve the problem of quantum reality and quantum measurements in a concrete, if unexpected, way. It also allows cosmologists to build theories of the universe without having to worry about the quantum reality of the whole thing. The idea of an alternative universe has even been applied to evolution. Some scientists have estimated that life is a highly improbable affair, that given the age of the earth and indeed the universe, it is highly unlikely that life could ever have evolved. Life may be highly improbable, argues physicist Henry Stapp, but there are a vast number of universes in which to take the chance. Our very existence implies that we live on one of those very rare, highly improbable realities in which life actually evolved. This argument, of the improbability of life, is not accepted by everyone. Belgian chemist Ilya Prigogine, for example, holds that it is based on outmoded assumptions and that the evolution of complex systems is not at all unlikely.

Some physicists, unconvinced by Bohr's refutation of the EPR paradox, attempted to press on with models of "local reality"—that is, theories in which electrons, photons, and atoms all have definite properties. A number of hidden-variable theories have been proposed in which the behavior of an electron, for example, is governed by totally real processes that take place at some subquantum scale. Surprisingly enough, Einstein was never convinced by these approaches, for he felt that they "get their results too cheap." His intuition was that some deeper theory was required, something more far-reaching than a naive return to classical physics at the subquantum level.

Most of these hidden-variable approaches involved the ideal of local reality, a concept that Einstein himself was loath to give up. In a local-reality theory, the properties of a quantum system—an electron, for example—are determined by the pushes and pulls, forces and interactions of some underlying reality. The theory may be sophisticated, but at base it is mechanical in the sense that things move only because they are pushed, pulled, or acted on by some force.

Although Einstein wanted to hang on to this idea of a local reality, he believed that this required some radical changes in our conception of reality. Einstein was perfectly willing to accept, for example, that our ideas of space, time, and causality may have to undergo a revolution. Toward the end of his life, he was even willing to allow for quantum indeterminism. Hidden variables may have appealed to other physicists, but for Einstein something totally new was required if local reality was to be retained at the atomic level.

But there is another possibility, that of a "nonlocal reality" in which distant parts of a system are correlated in new and mysterious ways. This correlation could not be achieved by any of the familiar energy-carrying forces and fields of conventional physics—otherwise we would be back with local reality. Rather, some mysterious sort of nonlocal connection would have to be used. Clearly such a subquantum account of a nonlocal reality would have to introduce a radical new idea and could not be entirely explained in terms of classical physics. David Bohm proposed such an idea of nonlocal reality with his notion of the "quantum potential" and "active information." We shall return to this idea of a nonlocal reality in the next chapter.

Multiple universes, quantum cuts, events determined by consciousness, hidden variables, and nonlocal interactions are just a few of the attempts physicists are making to escape the confines of quantum theory and its orthodox interpretation. In his book *Quantum Reality: Beyond the New Physics* (Garden City,

N.Y.: Anchor Press/Doubleday, 1985), Nick Herbert puts it bluntly, saying there is a twofold reality crisis in physics today: "1. There are too many of these quantum realities; 2. All of them without exception are preposterous."

Half a century ago, Bohr and his colleagues had met in Copenhagen in order to unify quantum physics in the face of dissension and the interpretation proposed by Erwin Schrödinger. Bohr had hoped to put an end to controversy, yet today things are in a worse situation, with a host of alternative interpretations of the nature of quantum reality.

The heart of physics is deeply confused, and the puzzle over the meaning of the quantum theory shows no signs of going away. Nevertheless, by the mid-1970s, with Niels Bohr dead, Léon Rosenfeld, now the leading proponent of the orthodox theory, still felt able to say, "The phrase *Copenhagen interpretation* is actually a misnomer, in the sense that there *is* only one interpretation of quantum mechanics. Bohr would rather say that quantum mechanics is a whole. . . . The misunderstandings that have been expressed so vociferously from various sides are based on a disregard of this circumstance."

But simply wishing it gone will not drive the controversy away. People are still puzzled about the quantum theory and its meaning. They demand to know the meaning of this mysterious quantum unreality. Is there any prospect of clarity within such confusion?

JOHN BELL

For decades physicists had argued about the meaning of the quantum theory but always in terms of thought experiments and abstract situations. No one believed that an actual experiment would come along and decide between some of these alternatives. This is why Bell's theorem appeared as such a shock to physicists. In the opinion of physicist Henry Stapp, "Bell's theorem is the most profound discovery of science."*

---

*H. Stapp, *Nuovo Cimento 40B*, 191 (1977).

Curiously enough, John Bell, who created this remarkable trick whereby nature is forced to reveal a deep truth about the nature of reality, was not a professional theoretical physicist concerned with the foundation of quantum theory. In fact, by profession he is a physicist concerned with the design of elementary-particle accelerators—a quantum engineer if you like. Red-bearded and soft-spoken, with a hint of a Northern Ireland accent, Bell is employed at CERN, the giant European elementary-particle physics laboratory outside Geneva, Switzerland. Talking to colleagues and fellow scientists, he displays a sense of humor coupled with an intellectual persistence that allows him to worry a physics question the way a dog worries a bone.

Yet even now, a quarter-century after his landmark discovery, Bell gives the impression that he is still puzzled by quantum theory, that he has still not teased out its essential subtlety. While most physicists can write down the equations and derivations of Bell's theorem, I wonder how many of them really understand its deeper meaning. As I spoke to Bell, I sometimes began to wonder whether he himself had fully plumbed its depths, or whether he would someday pull yet another trick out of his pocket.

UNWINDING THE PARADOX: FROM EPR TO JOHN BELL

The arguments that John Bell used to trick nature into revealing some of its secrets are both subtle and ingenious. Indeed, before we can fully understand just what Bell has achieved, it will be necessary to introduce and explain a number of new concepts. Bell's argument is a sophisticated and modified version of the EPR paradox. In Bell's reformulation, which began as a thought experiment and was only later reproduced in the laboratory, pairs of correlated particles are shot in opposite directions, *A* and *B*. Detectors at *A* and *B* record measurements made on these particles and, over many events, a set of correlations is observed between these two sets of records.

FIGURE 5-1

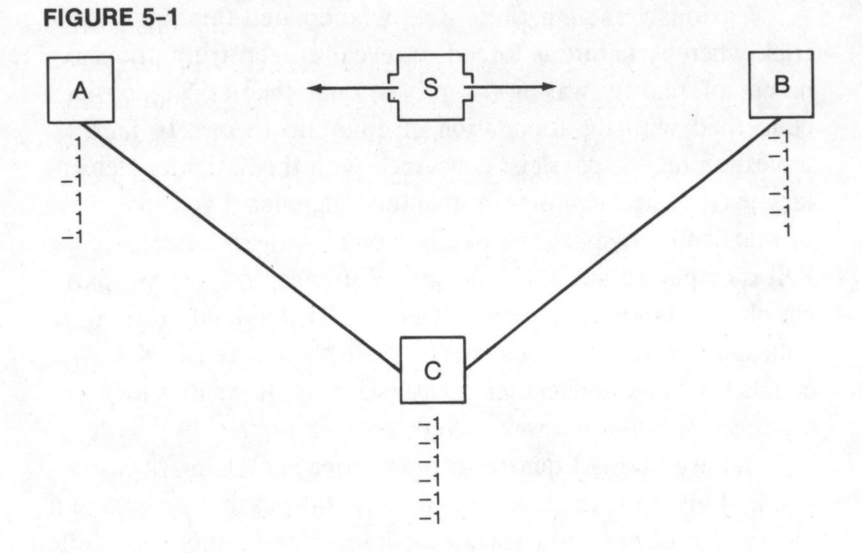

Bell's experiment. Correlated particles at *S* shoot off in opposite directions. They are detected at *A* and *B*. The outputs of the two detectors can then be compared at *C*. Note that there is no physical connection between the source and the two detectors.

Bell argues that two different theoretical predictions are possible for these correlations. One prediction is based on quantum theory; the other involves any theory that includes the notion of "local reality" (that is, any theory in which the quantum systems "possess" properties or could be said to carry cards of identity). The actual correlations predicted by these two alternative theoretical approaches turn out to be quite different, and Bell was therefore able to point to a way of experimentally deciding between them. In other words, Bell proposed a crucial test between the predictions of quantum theory and those of any theory based on the concept of local reality. The debate between Bohr and Einstein was about to be decided. The future of quantum reality hung in the balance. What would be the outcome?

The answer emerged only after some fine detective work. Bell's initial inspiration came from Einstein, who, for all his genius, had not been able to force nature into a corner with his EPR paradox. Demanding that nature should reveal a new secret was going to take something far more ingenious, and this is what Bell set out to do. As a young physics student, he was puzzled by quantum theory and could not take its results for granted, as other students seemed to do. He sympathized with Einstein's position and began to think about the great man's objections to Bohr's interpretation.

But after reading Max Born's philosophical treatise on the nature of the quantum theory, *Natural Philosophy of Cause and Chance*, Bell felt less secure about his reservations. After all, Born had pointed out that even if the quantum theory were superseded and new concepts developed, it would never be possible to return to classical physics. In the physics of the future, there would never be room for determinism or a reality composed of independent elements.

Bell's doubts were therefore put into abeyance until he came across some papers of David Bohm that inspired Bell to question again the whole meaning of the quantum theory. Writing in the mid-1950s, Bohm had introduced an alternative interpretation of the quantum theory, one involving a sort of nonlocal reality but nonetheless based on determinism—the sort of thing that other physicists had claimed was out of the question. In John Bell's words, "I saw the impossible done." Bohm's papers in effect opened a door for Bell, offering the possibility that Bohr was wrong and that a new sort of quantum realism could exist.

The new quantum realism hinted at by Bohm would not be a wholehearted return to classical ideas, since it involves radically new concepts at the atomic level. Here, at least, Bell and Bohm agreed with Bohr and Born: there could never be a total return to the old ideas.

But this was still speculating and building theoretical castles in the air. Was there some essential feature of nature that would direct Bell's thinking and suggest a new way in which quantum realism could be developed? Or was the very idea of subatomic reality to be eliminated from physics forever?

The answer emerged, in 1964, as a remarkable new idea called Bell's theorem. It showed that nature chooses between the old and new ways of thinking. The argument Bell developed is subtle, since nature could only be trapped into yielding its secret by a series of steps, each one calling upon some particularity of the quantum theory. In fact, Bell's argument begins with a variation on the original paradox of Einstein, Podolsky, and Rosen.

In an effort to sharpen the EPR paradox, David Bohm had earlier proposed a new thought experiment in which a correlated pair of electrons is used. Every electron has what is called a spin. But this "spin" is not the same as the spin of a spinning billiard ball. To start with, it can only have one of two values, which we call, by convention, "up" spin and "down" spin.

Bohr has warned us not to use such language or to use such ideas as an electron "having" a spin. In fact, it is better to phrase the end of the preceding paragraph in the following way: When a spin measurement is performed on the electron, only one of two results is possible, "up" or "down." In the language of Pascual Jordan, Max Born's collaborator, an electron does not so much "have" a spin as it is forced to adopt a particular value through a measurement.

It turns out that the particular value chosen is totally indeterminate. In any measurement in which the detector and source are properly oriented there is a fifty-fifty chance that an up spin is registered. Each measurement will involve pure chance, but over a long series of measurements, 50 percent will register up and 50 percent will register down. The following results are typical from a spin detector. The output has been designed to print a +1 every time an up electron is detected and a −1 when

FIGURE 5-2

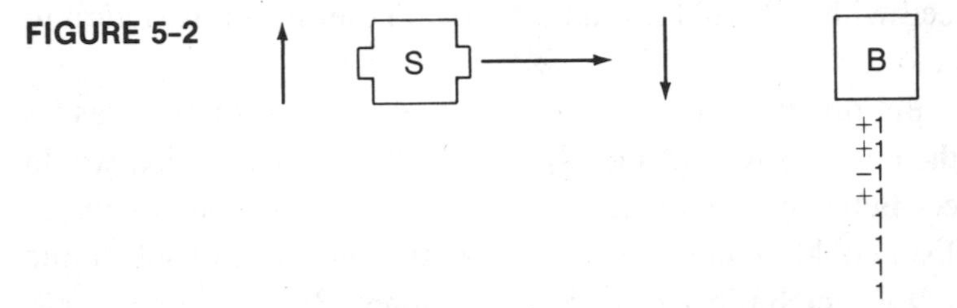

An electron shoots out from the source *S* and is detected at *B*. Spin "up" results are recorded as +1, spin "down" as −1. The output of *A* and of *B* is a purely random sequence.

a down electron registers. The sequence of +1s and −1s is totally random.

In Bohm's version of the EPR experiment, a pair of electrons, initially in a well-defined state, fly off in opposite directions. Measurements are then made on each electron at opposite ends of the laboratory.

At this point, it is important to stress the difference between Bohm's version of the experiment and that originally given by Einstein and his colleagues. In the original EPR thought experiment, all the measurements were carried out on one of the particles, and the other one is left alone. The goal of EPR was not utilitarian; it was not to gain any practical information about the second particle. Rather, EPR wanted to show that there was an objective meaning to speaking about the independent elements of reality for the second particle. Since the properties of the two particles are correlated, Einstein argued, by making, albeit, destructive measurements on one of them, it is possible to infer properties of the other. Bohr, as we have seen, objected to the whole trend of this argument and felt that it was based on a misunderstanding—that objects can have properties when we are not observing them.

In Bohm's version of the experiment, measurements are to be made on both particles. The goal is to confirm their actual correlation and to discover whether a quantum correlation in any way differs from a classical correlation. In essence, therefore, Bohm and EPR are doing something very different—although the actual experiments look similar, with correlated particles rushing off in opposite directions.

FIGURE 5-3

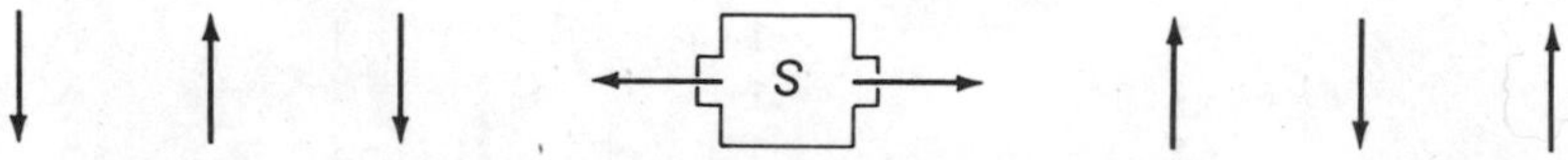

S

Electrons in the source are correlated in pairs so that if one has an up spin, its partner's spin will be down. The electrons then fly in opposite directions. While it is pure chance that an electron speeding to the right has an up spin, the direction of its spin remains correlated to that of its partner, which flies off to the left.

But just what is the correlation of these two particles? According to the mathematical formalism of quantum theory, it is possible for this pair to begin in a correlated way, so that if, in the future, one of them is measured in the up state, then the other will always be in the down state.

The diagram on page 94 shows the apparatus. The source *S* shoots off electrons in opposite directions, and detectors at *A* and *B* have paper tape outputs. The independent output of *A* and of *B* is purely random. In each case, there is an even chance of a +1 or a −1. Even after a long series of results, it will never be possible to predict whether the next event will yield an up or down—this is the essence of quantum indeterminism.

But what happens if we compare the two outputs? We discover 100 percent correlation. Whenever *A* records an up, *B* will register down, and vice versa. The following table compares the results of the two detectors. The left-hand column gives the random output of *A*, the right-hand column the random output of *B*. The correlation of the two results is shown in the middle column, which gives the product of multiplying the *A* and the *B* results. The answer is always −1. The average over many experiments is exactly −1.

FIGURE 5-4

Detector A	Combined Output C	Detector B
+1	−1	−1
−1	−1	+1
+1	−1	−1
+1	−1	−1
−1	−1	+1
+1	−1	−1
−1	−1	+1
−1	−1	+1
+1	−1	−1

The outputs of the two detectors *A* and *B* are totally random. These outputs are combined at *C*. Note that the results of *C* are perfectly correlated. Whenever an up electron registers at *A*, a down electron will be registered at *B*.

The perfect correlation between results at *A* and *B* means that a measurement at *A* makes it possible to predict exactly the result of a spin measurement at *B*. In terms of the EPR theory, one of the elements of the "independent reality of *B*" can be determined without having to interfere with *B* directly. At this stage, the experiment is simply a variation on the first step of the EPR paradox.

The results of Bohm's experiment can be accounted for by using conventional quantum theory, which predicts 100 percent correlation between *A* and *B* even though the measurements registered at *A* and at *B* are random. But is the same degree of correlation possible in a local reality theory in which electrons carry cards of identity, or is some new subatomic interaction needed to explain the perfect correlation of results? In fact, nothing new is required; local reality can also explain the Bohm experiment perfectly. There can be 100 percent correlations *without any interaction between* A *and* B!

To see how this is possible, take the case of the Bertlmann twins, who are identical in every respect except for their disagreement about hats. (The character of Dr. Bertlmann was introduced by John Bell. Here I have taken the liberty of turning him into twins and allowing him to wear a hat.) John Bertlmann picks up the first hat that comes to hand, while his brother, David, out of pure cussedness, chooses one of the opposite color. If John picks up a white hat, then David wears black. When John is wearing black, then David selects white.

The Bertlmanns work at opposite ends of town, and each morning you lie in wait outside John Bertlmann's office to spot which hat he is wearing. You can never predict the answer—it is pure chance—but once you have seen its color, you can phone your friend at the other end of town and predict exactly the color of David Bertlmann's hat. Once they have left home, John and David, at their opposite ends of town, no longer interact in any way, yet the color of their hats is always correlated.

The case of the Bertlmann twins indicates that in a classical world in which objects possess properties, just as the Bertlmanns wear colored hats, it is possible for distant objects to be perfectly correlated without needing to interact in any way or send any cross-checking signals to each other. In other words, just because we observe a correlation between two distant measurements, it does not necessarily mean that they are still interacting with each other—by means of signals or some curious nonlocal effects. Quantum theory predicts correlations, but so does any theory involving local reality.

In other words, we are no further ahead. The Bohm version of EPR is unable to distinguish between quantum theory and any "cards of identity" or hidden-variable version of an objective subatomic reality. Although Einstein and Bohr would differ about the interpretation of Bohm's result, they would certainly have agreed about its numerical predictions. Bohr would deny objective reality and cards of identity, Einstein would confirm it, but the supporter of orthodox quantum theory and the supporter of atomic reality would come up with the same answer: 100 percent correlation.

As a practical experiment, Bohm's experiment cannot decide between these two viewpoints. That would be left to John Bell, who would have to give Bohm's thought experiment a turn of the screw before it could distinguish between the two interpretations. And a turn is literally what it took—a rotation of one of the detectors.

The ingenious twist that Bell gave to Bohm's version of the EPR experiment was to change the relative orientations of the two spin detectors. Once this had been done, an absolute discontinuity could be established between the two worlds of local reality and quantum mechanics. At last it would be possible to decide which one applied in the atomic domain.

To understand why Bell's rotation of the spin detectors is so important, it is first necessary to learn a little more about electron spin. A spin detector could be thought of as a traffic signal that is placed at a fork in a road. This signal directs up electrons along one route and down electrons along another.

What happens if a second "traffic signal" is now placed

FIGURE 5-5

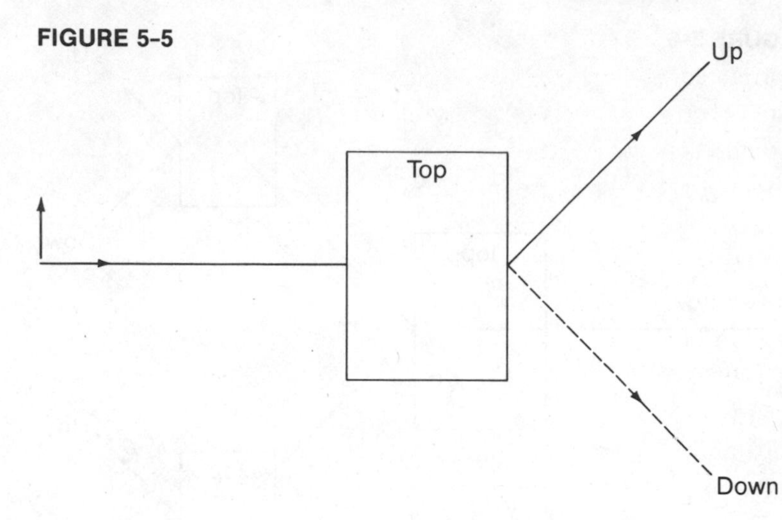

The spin detector switches up electrons along one track and down electrons along the other.

immediately after the first? If the second signal is located on the up road, then 100 percent of the traffic will continue to register up. No electrons will travel along the down path. In quantum mechanical terms, the result of the first measurement has forced the electron into a state in which its spin is up. Repeated measurements of this same property will always give the same answer: spin up.

Suppose, however, that the second detector is oriented at an angle to the first. It is supposed to distinguish between up and down electrons, but now electrons are oriented at a different angle to that detector. Quantum theory, however, demands only a simple up-or-down answer—a binary decision. It turns out that for each electron, there will be a given probability that it is forced into the up state or into the down state at the new orientation. (Again we are using a sort of shorthand to discuss this new configuration of the apparatus. According to Bohr and the Copenhagen interpretation, it is not permissible to talk of an electron "having" a spin. However, using a more careful choice of language, the same conclusions can be reached, but by a much longer route.)

FIGURE 5-6

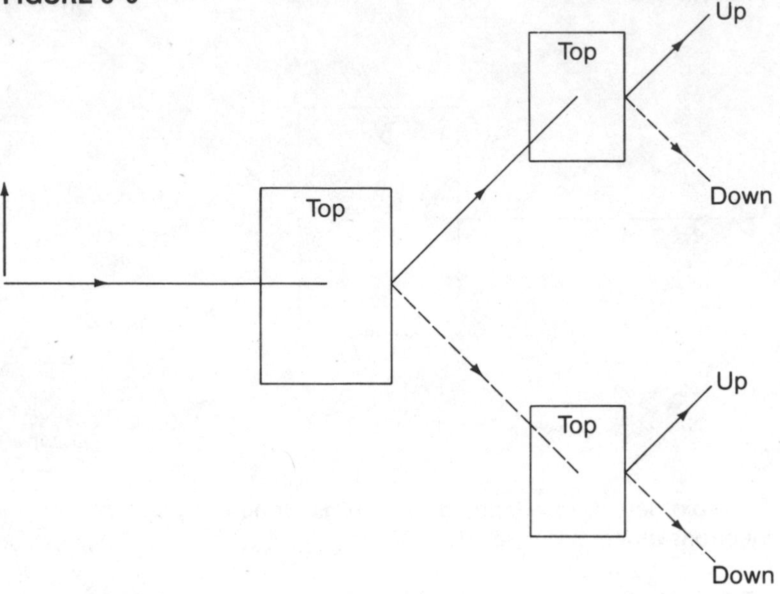

A second detector switches up electrons. No up electrons will take the second down path.

If the second detector has been given only a very small turn to take it out of alignment with the first, then 99 percent of electrons register as going along the new up path, and only 1 percent along the down. Twist the detector a little further, and 80 percent of electrons move onto the up path. With the detector twisted a full 180° (so that "up" becomes "down"), no electrons go along the up path, but all will go along the down. At an angle of 90°, 50 percent of electrons go along the up path, and 50 percent along the down. Hence, changing the orientation of the second detector with respect to the first changes the probability of an electron registering an up. The size of this probability depends upon the angle between the two detectors and can easily be calculated according to the rules of quantum theory.

Now let us return to John Bell's modification of David Bohm's experiment. As we have seen, with detectors *A* and *B* oriented parallel to each other, there is 100 percent correlation

FIGURE 5-7

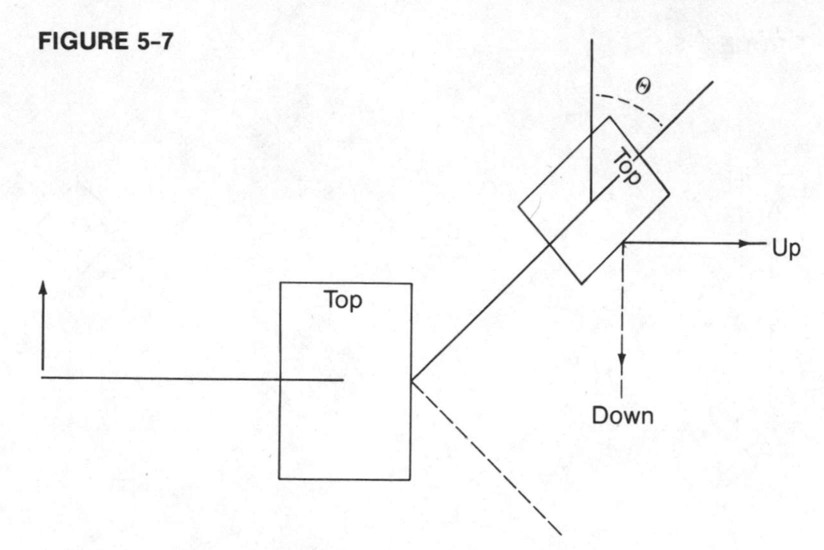

The second detector is tilted at an angle Θ to the first. Now an up electron encounters a tilted detector. Quantum theory dictates that, depending on Θ, a certain percentage of electrons take the up path and a certain percentage take the down. The smaller the angle Θ, the more chance an electron has to take the up path. With Θ at 90°, there is a fifty-fifty chance that an up electron will continue along the up path. With Θ at 180°, the second detector is upside down with respect to the first. Now all up electrons take the down path.

between the two sets of measurements—whenever an up is registered on one detector, a down will be registered on the other. Now give one of the detectors a very small rotation so that they are no longer perfectly oriented. This time, in the vast majority of cases, an up at *A* corresponds to a down at *B*. But occasionally an up at *A* gives an up at *B*. Rotate the detectors a little more, and the correlation decreases even more.

The following diagram gives the result of one such experiment. The left- and right-hand columns of +1s and −1s are distributed at random and correspond to the chance registrations of up and down spins at either detector. The third column, *C*, no longer shows perfect correlation. Not every entry is a −1; a few +1s now appear, so that the average value of *C* over very many events has changed from −1 to −0.9.

FIGURE 5-8

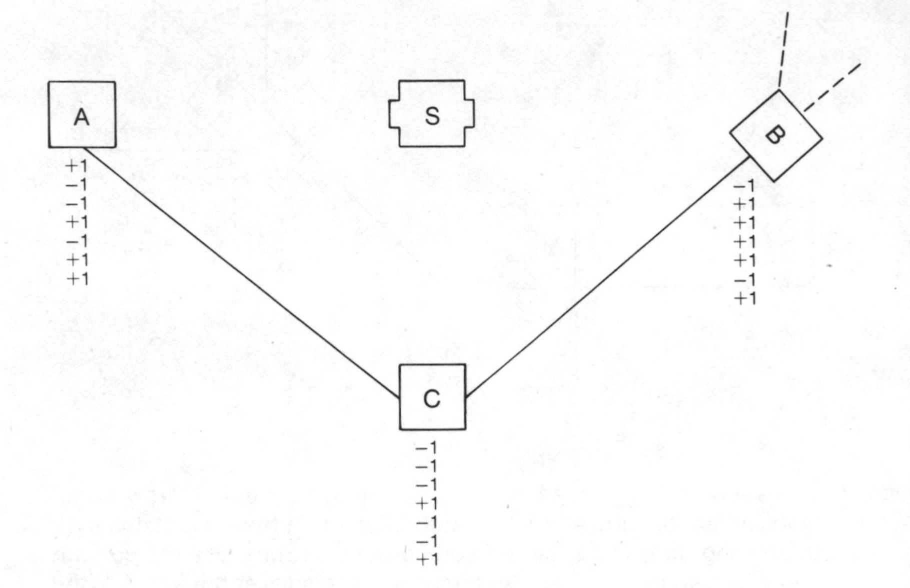

With detector *B* tilted at an angle Θ, not every up electron that hits the detector is registered as a +1. The output of *C* no longer shows perfect correlation. Some values of −1 occur. The average value of all the results at *C* has changed from −1 to −0.9.

FIGURE 5-9

Θ = 0°	Θ = 10°	Θ = 90°	Θ = 180°
−1	−1	−1	+1
−1	−1	+1	+1
−1	+1	+1	+1
−1	−1	−1	+1
−1	−1	−1	+1
−1	−1	−1	+1
−1	−1	+1	+1
−1	+1	−1	+1
−1	−1	+1	+1
−1	−1	+1	+1

The output of *C* for various angles Θ between the two detectors. With parallel detectors, there is perfect correlation.

Continue to rotate the detectors until they are at right angles with respect to each other. Now all correlation has vanished. When A registers an up, only 50 percent of events at B will be up, and 50 percent will be down. The average value of C is 0. If the rotation continues until one detector appears upside down with respect to the other, then correlation will be complete but reversed. If A registers up, then B will also register up. The value of C is now $+1$. By rotating one of the detectors with respect to the other, it is possible to vary the average value of C between $+1$ and -1.

The rules of quantum theory allow us to calculate the correlation probabilities for each particular orientation. In fact, the result is simple:

$$C = -\cos \Theta$$

where Θ is the angle between the detectors. When the Θ is 0, the detectors are parallel and the value of C is -1. When the two detectors are at right angles and $\Theta = 90°$, then C becomes 0; no correlation results. When one detector is upside down with respect to the other, $\Theta = 180°$, and $C = +1$. Experiments confirm the quantum theory's predictions.

But what about a local-reality theory? What sort of correlation would be established in a theory that allows electrons to "possess" properties like cards of identity? A total rejection of the Copenhagen interpretation implies that each electron has some sort of local reality that travels with it. Put another way, its properties can be defined by a set of variables, parameters, and interactions. These hidden variables or cards of identity determine the properties of an electron, including its spin, at every moment of time. On reaching the detector A, these variables interact with the internal working of the instrument to produce an effect—the registration of an up or down electron.

In the present experiment, the electrons are correlated in such a way that their spins are opposite. Their spin "cards of identity" are opposite, so that, in the case of parallel detectors, when one registers up, the other will register down. But what happens when the orientation of the detectors is changed? Suppose, for example, that down registers at *A*; this means that an upward-oriented electron reaches the detector *B*. But this time the detector is located at an angle to the up electron. Some as yet unknown interaction now takes place between the hidden variables determining this electron and the detector. Since the exact nature of the hypothetical hidden variables is unknown, it is not possible to predict the exact outcome of this interaction. All we can say is that for each down at *A*, not every local-reality electron registers up at *B*. Only a certain percentage of electrons registers up. Changing the orientation of the detectors will therefore break the 100 percent correlation—exactly the same trend as had been predicted in the quantum mechanical case!

If Einstein had been alive to reply to Bell's new proposal, what would he have said? The spin of the electron is a concrete reality, he would have argued, even if we may not understand its deeper nature. This spin is probably the manifestation of some underlying and as yet unknown subquantum processes. By altering the relative orientation of the spin detectors, the probability of registering an up or a down will change, and there will no longer be 100 percent correlation of results. Therefore, while disagreeing on interpretations, Bohr and Einstein would be in agreement on the trend of the numerical results: correlations fall off as the angle between the detectors increases.

As described earlier, quantum theory is able to produce an exact formula for this changing correlation: $C = -\cos \Theta$. What formula can be produced for the local-reality case? The answer is, of course, unknown, since the nature of the hidden variables has yet to be revealed. The supporters of a local-reality theory claim that once the detectors are turned out of alignment, their correlation will be less than 100 percent. This agrees with the qualitative predictions of quantum theory. The only problem is that, since the underlying variables that produce electron spin are not understood, no one can produce the exact numerical value of the new correlation. However, logic dictates that for a local reality, any correlation C must lie between -1 and $+1$, so, although there is no way of calculating its exact value, it can be expressed as an inequality. Of course, it is always possible that by pure chance this value will turn out to be the same as that predicted by quantum theory!

Suppose, for example, that a set of careful measurements are made and always come down in favor of the quantum mechanical correlation, $C = -\cos \Theta$. Does this mean that quantum theory has been confirmed? The odds appear to be in its favor, but there is always the chance that a rival local-reality theory could, by chance, come up with exactly the same numerical answer. Despite Bell's ingenious twist, we still have not tricked nature into revealing its secret.

Bell needed an experiment that would decide once and for all between two approaches to quantum reality, but since the exact new value of the local-reality correlation cannot be produced, there is no way of proving it is wrong! Some future local-reality theory could well devise a complicated formula that would produce, for a particular angle Θ, the same numerical correlation as the quantum theory. At this stage of the argument, there is no way of *proving* that such a formula can never exist so local-reality theory has not been logically disproved.

But Bell was not defeated. His next step was to logically prove that it would be impossible for such a "classical" formula to exist. Although the formula for correlations in a local reality is totally unknown, Bell was able to deduce that it must, nevertheless, have a certain overall form. His argument goes the following way.

If you believe in local reality, then the probability of an electron registering up or down at B will depend in a complicated and as yet unknown way on the interaction between the internal workings of that detector B and all the various hidden variables or cards of identity held by that electron.

Furthermore, no matter how local reality turns out to be constructed, one thing must be true about the situation. The result at B can have nothing to do with what happens at A. In other words, changing the orientation of *B* can have no effect on the probability of an up being registered at *A*. Thus, the two probabilities—of results at *A* and at *B*—must be totally independent of each other. What happens at one side of the laboratory can have no effect on what happens at the other. Therefore, this means that it should be possible in some way to split up the expression for $C_{\text{local reality}}$ so that it contains components that are determined by what happens around *A* and others determined by what happens around *B*.

While the details of this final expression for $C_{\text{local reality}}$ remain a mystery, Bell could be assured that its overall form, which contains local contributions from the region around *A* and from that around *B*, had to be different from the corresponding quantum expression. The quantum case, discussed earlier in this chapter, is:

$$C_{\text{quantum}} = -\cos \Theta$$

The simple term cos Θ cannot be broken apart; indeed, we learned in high school mathematics that it cannot be factored into two independent parts. It is evaluated knowing the orientations of one detector with respect to the other, no matter how far away those two detectors may be. While the local-reality result can be factored, or broken down into two parts, the quantum result is an indivisible whole. Thus, the two possible mathematical expressions for *C*—C_{quantum} and $C_{\text{local reality}}$—reflect the entirely different philosophies of quantum wholeness and of independent local realities.

Bell's deduction is a great advance in deciding between these two views of reality. We now know that the unknown local-reality formula must be entirely different from that given by quantum theory. But it is still not enough. The quantum and the local-reality approaches both give a numerical value for the correlation *C* that lies between +1 and −1. It still may be possible that by pure chance the two formulas, though totally different in form, will happen to give exactly the same numerical result for a particular experimental configuration. Local reality cannot yet be ruled out on experimental grounds.

A further step was needed—a second change in orientation. Bell showed that if the correlations *C* are recorded for one orientation, and the experiment is then repeated at a different orientation, it will be possible to decide between the two different predictions. The idea is to perform the experiment in the following way. First, establish the correlation with a particular orientation of the detectors. Then move the detectors to a different alignment, and measure the new correlations. In the local-reality case, it becomes possible to express the total correlation for both orientations, a new number $F_{\text{local reality}}$, in terms of the individual results. Since in a local-reality theory, whatever happens at *B* can have no effect on *A*, it is finally possible to distinguish between the two predictions.

Assuming local reality, in which electrons carry cards of identity, but in which detector *A* is so far from detector *B* that no mechanical influence, push, pull, force, or signal can pass between them, Bell was able to calculate the bounds on the correlation for the two-stage experiment:

$$-2 \leq F_{\text{local reality}} \leq +2$$

(Note that while in the previous version, one set of measurements was made on each detector to give a correlation *C* lying between +1 and −1, in this case two sets of measurements are made to give a correlation *F* lying between +2 and −2.) See "Bell's Derivation" on pages 110–112.

Of course, it is always possible to postulate that some signal can pass between *A* and *B*, but as we shall see, actual experimental arrangements will rule out this possibility. *A* and *B* can be arranged to be truly independent of any signal, force, field, or mechanical interaction.

Quantum theory predicts a different numerical result. The numerical expression for each state of the experiment cannot be factored or broken apart, and the result depends on the actual orientation of both detectors. When they are oriented at angles of

BELL'S DERIVATION

The experiment John Bell originally proposed involves determining, in two sets of measurements, the correlation between two distant detectors—the detectors being changed in orientation after the first measurement. Bell calculated the theoretical correlation in two different ways, first by using conventional quantum theory, and second by assuming only that reality is local. It turns out that an excess correlation is present in the quantum mechanical case over that for any local-reality theory.

The Local-Reality Case

P_a is the probability that a particular spin is detected by the detector *A*, located on the left of the apparatus. P_b is the corresponding probability for the right-hand detector. Neither P_a nor P_b can be larger than +1 or smaller than −1.

When the detectors are oriented parallel an "up" electron in detector *A* will register as +1. As the detector *A* changes its orientation, the value of P_a will decrease. If the detector *A*, on the left, is oriented at an angle Θ, the new case can be written as $P_a(\Theta)$. A similar expression applies for the detector *B*, on the right-hand side, oriented at an angle ϕ, for the case $P_b(\phi)$.

It is not possible to calculate the actual values of $P_a(\Theta)$ or $P_b(\phi)$, since the details of a hypothetical local-reality theory are unknown. What is certain, however, is that the value of $P_a(\Theta)$ must be totally independent of $P_b(\phi)$, for all effects are assumed to be local. In other words, the particular hidden variables, or other local effects, that determine the interaction of a particle with detector *A* cannot extend their influence across several meters to affect the detector *B*. In all cases, the correlations between the two detectors must be separable into a contribution from around detector *A* and one from around detector *B*. In the quantum mechanical case, however, things are very different, for the expression for correlation cannot be separated into two parts or factors.

To derive Bell's theorem—or the Bell inequality, as it is also called—a new expression is introduced, one that gives the correlation between the two detectors. $P(\Theta_a, \phi_b)$ is the probability that a particle will be registered at detector *A*, held at an angle Θ_a, while at the same time its companion is registered at *B*, held at an angle ϕ_b. Again, the value of this expression cannot be calculated; one merely knows that it must involve the product of two terms, one relating to detector *A*, the other to detector *B*. A local-reality theory dictates that $P(\Theta_a, \phi_b)$ must therefore lie between the values of +1 and −1. Nothing more can be said about it unless some complete local-reality theory is discovered. Note that $P(\Theta_a, \phi_b)$ is really the previous expression $C_{\text{local reality}}$, which gives the correlation between the two detectors *A* and *B*.

Likewise, $P(\Theta_a', \phi_b)$ is the probability that detector *A* registers a particle while oriented at a new angle Θ', while *B* is held at the old angle of ϕ.

$P(\Theta_a, \phi_b')$ gives the probability of a simultaneous measurement when *A* is held at its original angle of Θ, while *B* is rotated to a new position ϕ'.

$P(\Theta_a', \phi_b')$ gives the probability of both detectors registering a particle when both angles have been changed.

Bell then went on to derive the following relationship between the various probabilities. Its derivation in no way relies upon the physical details of the detectors. All that is assumed is that logic works and that what happens at one detector cannot have any effect on what happens at the other. Given that all the probabilites lie between −1 and +1, Bell showed that

$$P(\Theta_a, \phi_b) - P(\Theta_a, \phi_b') + P(\Theta_a', \phi_b) + P(\Theta_a', \phi_b') - P_a(\Theta') - P_b(\phi) \leq 0$$

Since $P_a(\Theta')$ and $P_b(\phi)$ lie between −1 and + 1, it is possible to rewrite the preceding expression in the following way:

$$-2 \leq P(\Theta_a, \phi_b) - P(\Theta_a, \phi_b') + P(\Theta_a', \phi_b) + P(\Theta_a', \phi_b') \leq 2$$

Writing the central term as $F_{\text{local reality}}$, Bell's expression becomes*:

$$-2 \leq F_{\text{local reality}} \leq 2$$

Quantum Mechanical Case

The formula for the quantum mechanical case is quite complicated and has to be worked out for each particular value of the angles Θ and ϕ. Unlike the local-reality case, it is not separable into a product of local contributions from the region of detectors *A* and *B*. It is a fully nonlocal expression. For the case of detectors oriented at angles of 45°, for example, $F_{\text{quantum theory}}$ is equal to 2.83, a number that clearly exceeds the upper limit of +2 demanded by any local-reality theory.

---

*Note that this derivation is a variation of the one first given by Bell.

45° to each other, for example, then this correlation is a maximum

$$F_{\text{quantum theory}} = 2.83$$

$F_{\text{quantum theory}}$ has other values depending on the orientation of the detectors. But certainly correlations like 2.83 are higher than a value of 2. In other words, the quantum world is more highly correlated than any world that depends on a local reality or locally operating hidden variables. Quantum theory predicts a sort of nonclassical correlation. The two approaches to reality have definite physical implications that differ in their numerical predictions, and these can be tested experimentally.

After decades of abstract debate, it is finally possible to decide between two approaches—orthodox quantum theory and any other interpretations based on the belief that electrons and other quantum particles have a definite local reality. In the case of local reality, spin correlations cannot be higher than +2, but in the former case, it is possible for a new sort of quantum correlation to manifest itself. Local reality implies that what happens at one end of the laboratory can have no effect on the other, unless actual physical signals and interactions are involved. By contrast, the formalism of quantum theory demonstrates an essential inseparability—the impossibility of factoring or breaking apart into independent processes the events at opposite ends of the laboratory. The two electrons in Bell's proposed experiment are correlated in a totally new way, a way that would be unimaginable in the world of local reality.

Thanks to John Bell's paper, published in 1964, it was possible to think of experiments that could decide between these alternatives. The meaning of quantum theory would be no longer a matter of academic debate but one of concrete measurements and precise calculations.

This is the result that Henry Stapp holds to be "the most profound discovery in science." In view of its significance, its conclusion is worth repeating. Quantum theory views any experiment, in Bohr's words, as an "undivided whole." In the case of Bell's experiment, the correlation between two detectors expresses this undivided wholeness in such a way that it is not possible to break up the mathematical expression for correlation into separate parts, one dealing with each detector. The resulting formula depends only on the relative orientations of the detectors—no matter how far apart they may be.

When it comes to a "realistic" view of quantum systems, in which electrons "possess" properties such as spin, the measurement at each detector must result from a complicated and as yet unknown process. Even if the hypothetical interaction is unknown, one thing is certain: the act of moving one detector can have no effect on the other, a great distance away. No matter how complicated the theory of local reality may be, the final expression for correlation can always be broken down into the product of two independent parts. A local-reality theory does not have this aspect of "quantum wholeness." No matter how ingenious a theory of "hidden variables" or local reality may be, there is no possibility of it giving that additional quantum correlation. Once reality has been broken down into its independent elements, then it has forever parted company with the more holistic quantum world.

The quantum correlations predicted by Bell's theorem are so new, so different from anything we have experienced in the world of classical physics, that it is difficult to illustrate or explain them. Since they lie totally beyond the world of signals, interactions, forces, fields, pushes, or pulls, none of this language can be used to discuss Bell's quantum correlation. But the following analogy may help.

In the early days of stereo, recording companies sometimes put out older monophonic discs that had been "electronically reprocessed to simulate stereo." These older discs simply did not contain the necessary acoustic information to produce a stereophonic effect. However, by means of some electronic filtering, it was possible to give the impression that violins came from the left speaker, cellos and double basses from the right, and violas from in between. This "electronically reprocessed mono" is a little like local reality. The sound coming from the left and the right speaker has a locally correlated order, which gives the impression that some sounds are on the left and some on the right, but it does not have that richness of stereo.

When you listen to the same piece of music in a true stereo recording, suddenly you have the impression of being inside a real concert hall. It is not simply that the violins appear to come from the left, but they sound as if they are being played in a real hall in which the violin sound is bouncing from walls and ceiling; there is a sense of space around the instruments—the whole recording has come alive. The reason is that the stereo recording contains additional information—it goes beyond the "local" information of the reprocessed mono disc. The sound reaching your ears contains the same information about wave interference as do the sound waves in a real concert hall. Rather than simply coming from the left speaker, the sound of the violins emerges from both speakers but in a more subtly correlated way so that the sound waves meet between the speakers and display complex interference patterns, simulating the sound of a concert hall.

The stereo recording stands for nonlocal, quantum correlations, the rich additional correlation that is not displayed in the local reality of the "electronically reprocessed" mono disc. A related effect is seen in watching one of those old 3-D movies. Without the glasses, the picture looks flat, but put the glasses on, and the whole scene leaps out at you.

EXPERIMENTAL RESULTS

Bell's remarkable paper "On the Einstein Podolsky Rosen Paradox" appeared in 1964. The next step was to plan an actual experiment. Since John Bell is a theoretician employed by CERN, the design and construction of apparatus had to be left to experimentalists. In 1969 a team of experimentalists and theoreticians, including John F. Clauser, Michael A. Horne, Abner Shimony, and Richard A. Holt, laid the plans for such a device. They also developed a slightly different version of the Bell formula that would be more appropriate to an actual experiment. It turns out that real laboratory detectors are not all that efficient—they sometimes miss registering a particle—and since correlation between detectors is at the heart of the experiment, it was necessary to refine Bell's argument to take this inefficiency into account.

By the 1970s, the first experimental results were being published. Instead of using correlated electrons, it proved simpler to use photons—quanta of light. Photons have something analogous to spin, called polarization, that can be easily detected, but in other respects the arguments and predictions are identical to the case of electron spin.

John Clauser, a young physicist working at Columbia University, had proposed to measure the polarization of twin photons using a Bell setup. At the time, he received little support for his suggestion and decided to move to Berkeley, where some of the necessary apparatus already existed. Working with Stuart J.

Freedman, he placed a sample of calcium in an oven, which would act as the source. Inside the hot oven, the calcium atoms were excited energetically until they shot off photons in opposite directions. Detectors at opposite ends of the apparatus, oriented first at 22½° and then at 67½° to each other, recorded the polarizations.

As to the final results, what looks straightforward on paper is more complicated in the laboratory. For example, no photon detector is 100 percent efficient, so many of the photons will never be registered. When the experimental results on correlation fall below 100 percent, does this actually mean that the quantum systems are themselves less correlated, or could it mean that the detectors are systematically missing some of the photons? To sort out how much of an effect results from the inefficiency of a detector and how much is really a deviation from quantum correlation, one must make a careful statistical analysis of very many experimental results. Nevertheless, in 200 hours of events, the experimental correlations were found to agree with those of quantum theory, clearly violating the alternative predictions made by local reality. Quantum theory had been confirmed.

This is the first of the experimental results that many physicists have considered to be the most significant in decades. In a sense, they transcend even quantum theory, for Bell's correlations are a fact about the nature of the world. Bell's derivation shows that no theory that clings to a local reality can be compatible with nature. Since the 1970s, other careful experiments have demonstrated the truth of his conclusion and have confirmed the predictions of quantum theory.

Suppose that at some time in the future, quantum theory becomes outmoded and is replaced by some other, deeper theory. One thing is now certain: that theory can never return to ideas of local reality. Whatever else it may be, the new theory must contain the same sorts of nonlocal correlations. Bell's result is profound. It has revealed an aspect of nature, one facet of reality that all future theories of physics must take into account.

But could the "quantum correlation" between two detectors be explained in some other way? Local-reality theory predicted that the number F cannot be larger than $+2$. Quantum theory, however, predicted an additional nonlocal correlation, a number F greater than $+2$. Experiments, like that of Clauser and Freedman, confirmed this additional connectedness. Since, however, the whole nature of reality is at stake, physicists must be certain that this higher correlation is an actual quantum mechanical effect and not some accident of the experimental setup itself. For example, is it possible that the two detectors were signaling to each other? A feedback in the electronics could produce a spurious effect that mimics an additional correlation.

The stakes on the Bell experiment are so high that each experiment has to be carefully scrutinized and criticized. All possible sources of error and misinterpretation must be isolated and identified. Throughout the 1970s, scientists in a number of laboratories, working with different experimental designs, tested Bell's predictions. The verdict still came down in favor of quantum theory.

Some of the most careful experiments were carried out by Alain Aspect and his colleagues at the University of Paris's Institute of Theoretical and Applied Optics in Orsay, near Paris. In this long series of experiments, calcium atoms are excited by lasers, which cause them to emit a pair of correlated photons. The photons then fly off in opposite directions, where they are detected. Each photon travels sixteen feet to its detector, making a total separation of thirty-two feet between detectors. The idea is to rule out any possibility of influence by one detector on the other. In a series of experiments over several years, each time using an improved design of apparatus, Aspect confirmed the predictions of quantum theory again and again.

In one set of experiments, published in 1982, the particular orientation of the detectors predicted a quantum correlation of 2.70. (As we have already seen, the local-reality theory demands that this correlation should lie between -2 and $+2$.) The actual experimental result came out as 2.697—again the nonclassical correlations of the quantum world were manifesting themselves. When all the errors in the experiment are computed, the experimental result lies between 2.682 and 2.712, clearly confirming the quantum predictions.

But even with the most careful experiment, a further loophole is possible. Alain Aspect had detected correlations between the measurements registered by his two detectors and ruled out any possibility of artificial correlations produced in, for example, the electronics. But could the two photons somehow be signaling to each other? Could there be some totally new and undiscovered effect whereby a photon at *A* would send a signal to *B* in order to trick *B*'s detector?

Aspect was able to avoid this objection by making the measurements so rapidly that there would not be enough time for any signal to travel the thirty-two feet between detectors before they had registered. Einstein's theory of relativity demands that no signal, no form of communication, can exceed the speed of light. That is why locating the detectors far apart and making measurements fast enough made it possible to rule out any influence from one part of the experiment on the other. Another loophole of local reality was closed.

But suppose that photons are indeed "realistic" and composed of much more complex subquantum objects. It is possible that, as they leave the source, their internal hidden variables are able to sense the actual arrangement of the whole apparatus and thereby collude to produce correlated results. The orientation of apparatus has been fixed in the laboratory for many minutes, which gives plenty of time for the calcium atoms and lasers to gather "information" about the orientation of the detectors and somehow transmit that information to the photons as they are created. In this way, two "realistic photons" may collude to fool Aspect's detectors into believing that additional quantum correlations exist.

This may sound like farfetched science fiction, but its possibility has to be ruled out. What is at stake is atomic realism, something that physicists have believed in for over two hundred years. Any loophole that the human imagination can conceive has to be considered.

If two realistic photons happen to know about the orientation of the apparatus before they are shot to opposite ends of the laboratory, it is possible, as with the Bertlmann twins and their different-colored hats, for them to mark their cards of identity accordingly. The only way to rule out this possibility of collusion is to quickly change the orientation of the detectors while the photons are actually in flight. The idea was to change the orientation of *A* so fast that a signal would not have time to catch up with the other photon before it was registered at *B*.

The two detectors are located twenty-six feet apart, and it takes light 40 nanoseconds to travel that distance. (One nanosecond is 0.000 000 001 second.) No physical signal could travel that distance any faster because no signal or interaction can travel faster than the speed of light. Aspect's idea was to change the orientation of the detectors *after* the photons had begun to fly apart. Making that change within 10 nanoseconds would ensure there was too little time for any new signal to pass between them.

In this way, Alain Aspect had achieved what is called Einstein separability—two distant events that happen so quickly that there is not enough time to connect them by any signal or interaction. These events—the registration of the two detectors—are absolutely separated according to Einstein's relativity. There is no way that one detector can know about the other until the measurements have been registered; all possibility of hidden-variable collusion has been ruled out.

How is it possible to switch detectors that fast? Clearly no mechanical rearrangement could take place within a hundred millionths of a second! Aspect's solution was ingenious. He used the surface of a liquid to act as a mirror and reflect the photon into a detector. Imagine an ordinary mirror that bounces light from a lamp onto a detector. Now tilt the mirror so as to reflect the light in a new direction, onto a second detector. By rotating the mirror back and forth between these two positions, you are

FIGURE 5–10

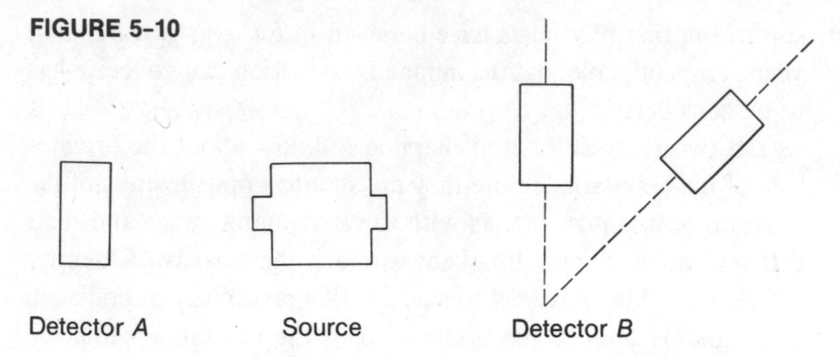

In Aspect's experiments, the orientation of the detector on the right must be changed so rapidly that no signal can pass between *B* and *A*.

effectively switching the light from one detector to another.

Although tilting a mirror is much faster than physically rotating a detector, it still takes too much time. Aspect needed to discover a way of directing the beam of light using a liquid. The key was to set up a very fast vibration of the liquid—an ultrasonic sound wave—producing a complicated internal pattern of waves that had the effect of directing light into one or the other detector. Using ultrasonic vibrations, it was possible to switch a photon between detectors in only ten nanoseconds.

Most physicists are well satisfied that quantum theory has been confirmed and local reality has been ruled out by Aspect's experiment. But there are still ingenious critics who want to close every possible loophole of a conspiracy theory on the part of a local-reality nature. Despite all of Aspect's experimental precautions, is it still possible to imagine a local-reality world in which nature conspires to push correlations above a value of +2, toward the quantum value?

One possibility for a hidden-variable fraud lies in the way in which the actual switching between detectors is controlled. The acoustical wave in a liquid, which deflects a photon into one detector or the other, is controlled by the electronics of Aspect's equipment. Is it possible that the photons could "know" about the inner details of this electronic control? Even when a random-number generator is used to trip the vibrations, it is still possible that this generator is not truly random but operates according to a complicated electronic program. Thus, a bizarre form of local-reality conspiracy, although it sounds extremely farfetched, is still possible.

Aspect's latest proposal is to use the light from a distant star to trigger vibrations in the liquid and hence switch photons between the detectors. Surely a local-reality conspiracy cannot stretch across the universe. The light that reaches earth today left the distant quasar many millions of years ago. At the instant of its creation, surely it could not have anticipated that on some remote planet a French scientist would choose a moment in the distant future to carry out a quantum measurement. Processes taking place at the ends of the galaxy can have no connection with what happens millions of years into the future within a Paris laboratory—unless, that is, the whole universe is itself some giant intelligence that conspires to preserve the illusion of a quantum nonlocality.

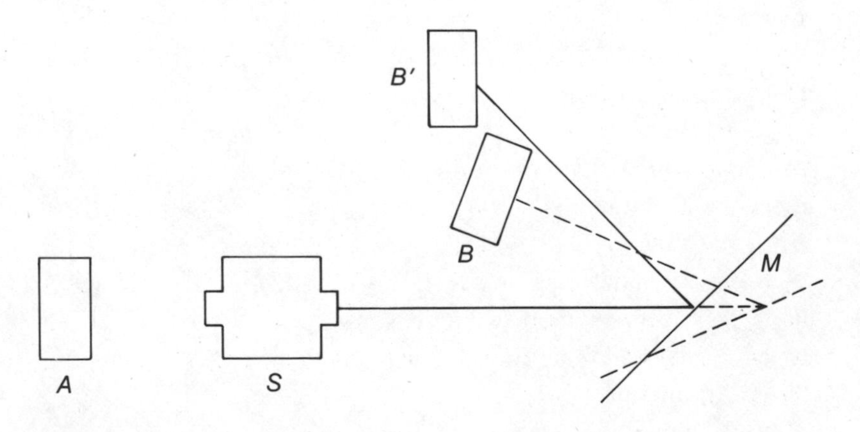

Detector *B* is not physically moved. Instead, a photon is reflected into one of two detectors, held at different angles, by the mirror *M*.

Scientists agree that the many different experiments employed to test John Bell's predictions come down firmly in favor of orthodox quantum theory. Einstein's position is now untenable—local reality cannot be retained, and physicists must agree that the quantum world is stranger than anyone imagined. The implications and interpretations of Bell's result are therefore far-reaching. They involve concepts like nonlocality and quantum wholeness. They have also given rise to a number of last-ditch attempts to hang onto subatomic reality, such as curious super-luminal connections.

6
IMPLICATIONS AND EXPLICATIONS

The implications of Bell's theorem are staggering. Had Bohr and Einstein known about Bell's results, their dialogues would have been very different and may well have profoundly influenced the future of physics. Einstein for his part would have been forced to rethink his objections to the quantum theory and, indeed, the very basis of his whole philosophy of realism. Einstein had asked, "Is the moon really there when no one looks at it?" Thanks to Bell's theorem, the implications of his questions are becoming stranger than anyone suspected. It turns out that in a sense we are all connected to the moon, even if we do not happen to be looking at it.

Indeterminism and quantum uncertainty, while they posed a difficulty for Einstein, were never a major stumbling block. Indeed, toward the end of his life, Einstein was willing to admit that physicists may be forced to give up causality and determinism at the subatomic level. But he was not ready to abandon the central idea of an independent reality. Einstein believed that there must still be an objective meaning to the reality of electrons and photons, even if totally new concepts have to be created in order to talk about them.

But what exactly is reality? Bohr said that it is a word in our language and that, as with any other word, physicists must learn to use it properly. Bell's theorem shows that if physics wants to hang on to that word *reality*, then it must be used in totally new ways. Reality can no longer be restricted to a purely "local" meaning, for the nonlocal implications of Bell's theorem show that what happens in one region of space is correlated to other distant regions in the universe. In the future, *reality* must also be used in a nonlocal sense, suggesting that our concepts of space and time may need to be transcended. The implications of Bell's theorem may well demand an even greater revolution in thought than what had first been expected from relativity and quantum theory.

Of course, there is still a way out of this. Bohr's own position was to deny that quantum reality can be talked about in an unambiguous fashion. All that is certain are the results of measurements made with large-scale apparatus in the laboratory. To try to go beyond these measurements and speculate about the reality of the quantum world is a fruitless occupation. The choice before us is either to abandon any hope of knowing the nature of quantum reality or to accept a nonlocal universe.

If Einstein and Bohr were alive today, what would have been their reaction to Bell's theorem? Perhaps the demand of a nonlocal reality demonstrated by Alain Aspect's experiment would have forced Bohr and Einstein to face the gap between them. Einstein was a deep and committed thinker whose intellectual honesty would not have avoided the implications of Aspect's result. In attempting to reconcile a nonlocal world with the insights into space and time uncovered by the theory of relativity, Einstein might well have given physics a great new surge forward. Niels Bohr, for his part, had already speculated that our current theories of space and time may turn out to be inappropriate at the atomic level, so some radically new ideas are required. These new ideas may evolve out of an attempt to reconcile quantum theory, relativity, and the ideas of nonlocality inherent in Bell's theorem.

But Bohr and Einstein are no longer with us, and neither is Wolfgang Pauli, that iconoclast and merciless critic of all ideas that are slipshod and half-baked. How do contemporary physicists react to the implications of Bell's theorem?

A QUANTUM NONEVENT?

It is possible to take the view that Bell's result is a nonevent and that nothing new has been revealed about the world. Quantum theory predicts a certain degree of correlation between distant detectors, and as everyone expected, that correlation is confirmed experimentally. Mainstream physicists never believed that quantum theory was in any serious danger, and therefore no new insights into its formalism have been gained. In the world of hard-nosed physics, there is no room for endless debates about the nature of reality. All that matters are experiments and calculations, and if the calculations always agree with the data, then everyone is happy. Bell's theorem has done nothing to disturb this position—so why all the fuss?

According to this point of view, the best that can be said about Bell's theorem is that it has put an end to half-baked theories about hidden variables. The bad news is that it has sparked a series of conferences and papers in which physicists wax metaphysical and speculate on the meaning and interpretation of Bell's result. As far as hard-nosed physicists are concerned, scientists should stop talking philosophy and get on with their real work, which involves solving problems, making calculations, refining theories, and doing careful experiments. Metaphysics has no value in physics, they would say—it is the preoccupation of elderly scientists whose creative life is over and who enjoy puttering around the philosophy of science. As to that talk about nonlocality, no one needs to bother about nonlocality to calculate the energy levels of an atom or the electronic structure of a metal.

Nonlocality

Some commentators have speculated that mainstream physics would have been far happier if Bell's theorem had never been published. His curious quantum correlations may well have spelled the end to hidden variables and local reality, but in putting them forth, Bell had allowed something new and objectionable to creep in—the idea of nonlocality itself.

In fact, the notion of nonlocality had always been present in quantum theory. Bohr talked about the undivided wholeness of a quantum experiment and pointed out that the quantum system cannot be separated from the laboratory equipment that measures it. It is true that physicists may not have been forced to face the issue until John Bell came along. Nonlocality is the black sheep of the quantum family; no one can deny it exists, but it is in bad taste to draw attention to its behavior. Bell had brought the scandal of nonlocality out in the open.

Why should the idea of nonlocality be so shocking? The reason is that, for hundreds of years, scientists had regarded the world as a great mechanism in which the wheels of the universe are turned by a variety of pushes and pulls. If anything moved, it was because something else acted on it. Everything that happened in nature could be understood in a purely local way. Within such a scheme of things, nonlocality would be external, mysterious, and unnecessary.

For much of his life, the great Isaac Newton was troubled by the meaning of gravity. His theory of mechanics could account for the motion of the moon and the fall of an apple, yet the force of gravity had something of the character of an "action at a distance." Nothing lies between the earth and moon except empty space. Nevertheless, gravity acts across this nothingness to exert its pull on the moon. There was almost something magical about such a force that appeared to operate by remote control. The idea of action at a distance was abhorrent to Newton and his contemporaries. Today nonlocality is viewed, in some quarters, in a similar light.

Einstein, of course, was able to explain the nature of gravity in terms of the curvature of space—that is, in a purely local way, since each tiny local region of space is curved. James Clerk Maxwell also was able to explain the forces of electricity and magnetism in terms of a new concept called a "field"—but again fields could be defined locally at each point in space. All forces, all pushes and pulls of physics, although acting across great distances, could then be explained in terms of fields or space-time curvatures that are well defined from point to point. In addition, these forces and interactions fall off with distance. The farther you separate two objects, the smaller is the interaction between them.

So physics had been wedded to locality, to systems that were well defined within their own particular regions of space and time, and to interactions that involved the transmissions of physical energy across space. The idea of nonlocality appears to deny this most basic fact of everyday experience, for it suggests that distant systems can be connected in a totally new way—a way in which distance no longer seems to matter. It is this proposal that is profoundly disturbing to many physicists.

A SUPERLUMINAL REALITY?

To those few die-hard physicists who had never accepted the quantum theory and instead wanted to hold onto the idea of locality and "independent elements of reality," Bell's theorem came as a great blow. How was it possible to hang on to the ideas of a local reality except by finding a fundamental loophole in Bell's argument or in the experiments themselves?

Bell's theorem shows that orthodox quantum theory predicts a correlation between distant objects that cannot be explained by any appeal to local reality. Experiments by Alain Aspect and others confirm this result. Moreover, Aspect's measurements were made in such a way that no signal moving at the speed of light could connect the two detectors and cause them to become correlated.

Today the only possibility of hanging on to a local-reality theory is to suppose that the Bell correlations are somehow the result of a physical interaction or signal that passes between the detectors at a speed that is faster than light! But, as we have seen, such superluminal connections would violate Einstein's theory of relativity.

In fact, superluminal signals lead to all the problems and paradoxes that Einstein had pointed out at the turn of the century. Relativity teaches that the appearance of events depends on the motion of the observer. One observer may see two events as happening simultaneously, while another sees them as separated by a short interval of time. But despite differences in appearance, one thing is certain: the events themselves cannot violate the principle of causality. In our universe, a bell cannot sound until it has been struck, and ripples do not spread in a pond before a stone hits the water. But once superluminal interactions are allowed, then effects can occur *before* the events that caused them!

In a superluminal world, light can flood a room before the switch is thrown. According to some observers, the signal to fire a rocket will be received before the order is given. Introducing superluminal signals allows paradox into the universe. It may save local reality but only at the expense of throwing away causality and relativity. In an attempt to save the bathwater, the baby would be thrown down the drain!

Since most physicists agree that faster-than-light signals and interactions make no sense, they do not take such proposals seriously. They may represent a way of holding on to local reality, but the cost involved is too great. Nevertheless, there are some thinkers who are quite willing to entertain the idea of superluminal connections; in fact, they welcome them and have even called on Bell's theorem to justify these assumptions. Scientists who are interested in paranormal phenomena, such as remote viewing, ESP, and the action of the mind on material objects, have attempted to explain these events in mechanistic terms. Requiring a new form of interaction between the brain and the material world, and between brains and brains, they have proposed an instantaneous connection over long distances. What better avenue of exploration of the paranormal than to invoke Bell's theorem and appeal to superluminal interactions? If two quantum systems can be correlated at a distance, then why not two brains?

There are even hints that the U.S. military is getting into the act. Writing in *Physics Today*, David Mermin quotes from the text of a letter sent from a California think tank to the Under-Secretary of Defense for Research and Engineering:

> If in fact we can control the faster-than-light nonlocal effect, it would bc possiblc . . . to makc an untappable and unjammable command-control-communication system at very high bit rates for use in the submarine fleet. The important point is that since there is no ordinary electromagnetic signal linking the encoder with the decoder in such a hypothetical system, there is nothing for the enemy to tap or jam.*

One shudders to think what would happen if the U.S. fleet's superluminal signal ever became confused with that used in a séance of spoon benders!

Physicists have been able to prove that, according to the rules of quantum theory, the Bell correlations cannot be used to send signals from one part of the universe to another. Two electrons may indeed be correlated, but conventional quantum theory dictates that such correlations cannot be used to transmit

---

*Since that letter was written, two physicists, Charles Bennett and John Smolin, at IBM's Watson Laboratory in Yorktown Heights, N.Y., have announced an absolutely secure "quantum communications device" called Quantum Public Key Distribution (QPKD). In essence, the device establishes a quantum connection between transmitter and receiver, using very weak light. Any attempt to intercept and overhear the message results in an uncontrollable quantum disturbance of the signal and thereby destroys information. While it does not employ superluminal signals, the QPKD is using the basic quantum property that noncommuting observables cannot be measured simultaneously, so that any attempt to "measure" part of the message has the effect of destroying it.

thoughts and impressions between two brains, or to send signals from one spy to another.

The Bell correlation is extraordinarily delicate. If it were to be used as a signaling device, then a message would have to be superposed on the correlation—for example, to modulate it or to switch it off and on. But any attempt to interfere with or disturb the correlation will destroy it. Bell's theorem offers no easy way out for nonlocal communications.

QUANTUM WHOLENESS?

Leaving superluminal signals and die-hard local realists aside, what is the true significance of Bell's theorem? The careful experiments carried out by Alain Aspect and others show that quantum systems are correlated in ways that defy explanation in terms of any connections, interactions, fields, pushes, or pulls that would have any meaning in conventional physics. There is a "wholeness" about the quantum world that is totally alien to older, mechanistic ways of thinking.

Quantum wholeness is not the result of any force, signal, or interaction. No force field or physical signal connects the two quantum systems; no interaction passes between them. In fact, quantum correlations could be said to be instantaneous, since Bell's theoretical result dictates the presence of a correlation even when two distant measurements are carried out simultaneously. Although the idea of an *instantaneous* connection has not been proved experimentally, it has been supported by experiments in which measurements are carried out so rapidly that no signal traveling at the speed of light could connect them.

But how does it make sense to talk about a correlation between distant parts of the universe when no interaction connects them? How can there be a connection between remote events when no signal can convey information from one system to the other, or the one can never learn what the other experiences? The idea seems paradoxical. Indeed, it lies beyond any normal explanation for the very reason that, as soon as we think about

Bell's correlations, we find ourselves slipping into patterns of thought that are based on our everyday conception of the world. Correlations, in our usual way of thinking, imply a connection. So how can a correlation exist when nothing passes between the systems? How can things be connected if nothing material—indeed, no sort of field—lies between them?

Bell and others have referred to quantum correlations as nonlocal effects. But what exactly is nonlocality? We have agreed to throw out local reality, but in its place this new word, nonlocality, has appeared. What does it mean?

Trying to understand nonlocality is like confronting the Zen koan "Do not think of a monkey." As soon as the novice monk tries *not* to think of a monkey, the idea of a monkey jumps into the mind. How can one *not* think of a monkey without bringing in the very thing one does not wish to think about? John Bell is telling us, "If you want to understand the quantum world, do not think of locality!" But as soon as we try to think of nonlocality, we find ourselves using the language of locality.

As we try to come to terms with nonlocality, we find ourselves using words like *in contact*, *separate*, and *long distance*, which are based on local reality! Things are in contact when they touch. They are far apart when it takes a long measuring tape to connect them or when it takes a long time for a signal to travel between them. Our feeling of distance is based on the idea of connections involving a large number of small local steps. Distance has meaning when we measure it using tape, radar, or laser beams, that is, when we span and cover that distance with some physical object, signal, or field.

Any attempt to pin down nonlocality seems to take us back to locality. As soon as we try to discuss the significance of Bell's result, we find ourselves bound up in ideas of space that are inexorably entwined with the concepts of local reality. Could it be that our notions of space and of time are inadequate within the quantum world? Does nonlocality demand a new language?

We have discovered that as soon as we start to talk about quantum nonlocality, we find ourselves using words like *distance*, *contact*, *separation*, *far*, *near*, *signal*, *interaction*, *measure*. These words are all grounded in our everyday experience of the world and are intimately tied to the idea of locality. How can we discuss nonlocality when the very language we use is geared to locality?

Bohr had already anticipated this problem when he remarked to his colleague Aage Petersen, "We are suspended in language in such a way that we cannot say what is up and what is down." This problem of language was, in Bohr's opinion, crucial to the problem of understanding quantum theory. As a mathematical formalism, quantum theory is a series of equations and calculations, patterns of symbols on a piece of paper. It is only when we begin to talk about these equations that their meaning is revealed. "What does the symbol Ψ mean?" we ask. "How does Ψ change with time?"

When physicists think and talk about the meaning of their equations, they are of necessity forced to use the same everyday language as everyone else—spiced up, of course, with some technical terms. In other words, they gather around a blackboard and talk in English, Russian, French, or some other language, but this language has been conditioned by hundreds of years of looking at the world in a particular way. It has been conditioned by Newtonian-Cartesian physics and by the fact that we live at a scale of size and energy in which reality is taken for granted. In our everyday experience, objects do not jump from particles into waves, people cannot go through two different doors at once, and distant relatives use the telephone when they wish to communicate with each other. Language, through its very structure, reflects this large-scale "classical" world.

On the one hand, our minds try to probe the ephemeral reality of the quantum world; on the other, we talk, think, and act in a language adapted for discussing trees, rocks, and automobiles—as well as poetry and emotions. As Bohr pointed out, the only way we can discuss the meaning of the quantum formalism is with a language that has been adapted to the requirements of the world of classical physics. We are suspended in this language so that we do not know which way is up and which is down.

Attempting to discuss the nonlocality of Bell's correlations brings us face to face with this paradox.

Is there no way out? In Bohr's opinion, we are stuck with the problem: "All experience must ultimately be expressed in terms of classical concepts. The account of the experimental recording of observations must be given in plain language, suitably supplemented by technical physical terminology. . . . In this sense the language of Newton and Maxwell will remain the language of physicists for all time." We are asked to think of nonlocality, yet as soon as we frame our thoughts, the language of space, time, and locality dances inside our heads.

Images of Nonlocality

Is language therefore forever limited and the world of nonlocality closed to us? Poets and writers are never afraid to enter a new dimension of experience or perception and bend language to their will. There may indeed be a truly creative way of using language in order to come to grips with the "reality" of the quantum world and with the meaning of nonlocality—without, that is, getting caught in the traps of our current conventional use of language.

As a starting point, it is helpful to think of space, of locality and nonlocality, in terms of a series of illustrations and metaphors. The idea is not to build up a feeling for nonlocality in any precise sense, but more to give an impression of what it may mean. The result will be like a series of paintings in which the artist has not attempted to be "realistic" in a photographic sense but has used impressionism, cubism, and expressionism to portray different aspects of the one scene. By creating a montage of metaphors, it may be possible to gain an intuitive understanding of the meaning of nonlocal reality. Only by breaking the hypnotic hold of locality and Cartesian space will we open our minds to the possibility of nonlocality.

Take, as a first illustration, a drop of ink placed on a tangle of string. When the string is unwound, the ink dots become very far apart. Measured along the string itself, the distance between the dots is several feet, but across the tangle they were only a few millimeters apart. The two points are both close together and far apart. To put it another way, measured with respect to a linear unfolded order, the points are a great distance from each other. But with respect to the tangled enfolded order, they are in contact.

Does an analogous order exist for space? Could there be two measures, one an explicit, Cartesian order in which two events may be far apart, the other an enfolded order in which the same events appear close together? In fact, the idea of folding is a good one. A cook, for example, talks of "folding" an egg into cake batter. In one sense, the batter is folded around the egg, while in the other, the egg is folded around the batter. The egg is therefore contained within the batter, while at the same time the batter is contained within the egg. Both are enfolded within the other. By this process, threads of egg may be drawn out a great distance yet constantly folded back on each other. The ends of the threads are both far apart and close together.

In two simple examples from our everyday experience, we have moved far beyond the conventional geometric concepts of distance and proximity, inside and outside. Could space at the quantum level have this nature? Does an electron lie inside the experimental apparatus, or does it enfold it? Does the electron pass through two different slits, or are apparatus and electron enfolded each within the other? Could it be that several different orders of space are simultaneously possible, each one revealing a different aspect of reality?

A painting is a two-dimensional portrayal of a solid, three-dimensional reality. As your eyes move around the painting, you feel that you are moving within a real three-dimensional space. A fraction of an inch to the left, and you are looking at a human ear; a fraction to the right, and you are looking at a mountain in the far distance. Small movements across the surface of the painting represent great leaps through three-dimensional space. Distance can be stretched in one direction and compressed in another. In a painting by Cézanne, for example, planes simultaneously recede and advance in a dynamic way, and boundaries appear simultaneously to be at different distances from the viewer. By adopting a whole series of strategies—or schema, as the art historian Ernst Gombrich calls them—painters have learned to portray on a two-dimensional surface all the complexities of a three-dimensional space.

A holograph is another way in which information about a three-dimensional space is folded over a two-dimensional surface. Whereas in a photograph each part of the scene is associated with a definite part of the photograph, in a holograph the scene is enfolded over the whole holograph so that each part of the holograph contains all of the scene, and each part of the scene is contained in every part of the holograph. Illuminate only a small part of the holograph—or break off a small piece and illuminate it—and one still sees the complete scene, albeit with more "noise" and less resolution. This means that distant parts of the scene are folded together and are in contact within the holograph, while at the same time neighboring points are spread out over the entire holographic plate. Just as with kneaded bread dough, space has been both stretched and enfolded. Events are mutually contained within each other so that the distinction between inside and outside is lost. Again our familiar conception of space and distance is turned on its head, and we see ways in which the mind could accommodate the notion of nonlocality.

Geometrodynamics—Of Black Holes and Wormholes

Despite the revolutions of quantum theory and general relativity, physics has continued, until recently, to use the same mathematical tools that were created by Descartes and Newton. The geometry taught in schools involves measuring lengths, areas, volumes, and angles. In a more sophisticated form, this sort of geometry underlies much of the mathematics that is used to describe space-time and the paths of particles. Even when Einstein made his break with the flat geometry of Euclid in favor of a space-time full of curves and bumps, he used the same underlying geometric concepts of distance and angle.

General relativity relies upon a space-time whose geometry is really an extension of the Euclidean geometry taught in schools, modified to allow for the curvature of space by matter and energy. But is there a generalization of geometry that lies beyond even this—one that is deeper than geometry based upon distances, angles, and shapes? It turns out that a branch of mathematics called *topology* is more general than even the non-Euclidean geometry used by Einstein. If physics wants to make a break from the constraints of locality and explore some of the implications of nonlocality, then a study of topology may well be the starting point.

Topology is concerned with space as connections and intersections, with the distinctions between the inside and outside of something. Since topology does not deal with distances or with the distinctions between different shapes, it is often spoken of as geometry on a rubber sheet. It is a study of things that can be stretched, twisted, compressed, even torn and rejoined. Drawn on a sheet of stretching rubber, a circle, triangle, and square become indistinguishable. Yet since all are closed figures, they are different from a line or an intersection of rings. According to topology, a ball, pyramid, and cube are equivalent, since one can be transformed into the other, but they are different from a doughnut or cup, which contain a hole. Objects with holes in them can never be transformed by simple stretching and deforming into objects without holes.

Since stretching and deformation are unlimited in topology, the whole idea of distance and angle does not apply. Is this perhaps the place to start to think about a space that could be nonlocal? Certainly a number of physicists have explored the ideas of topology. Indeed, the recent explosion of interest in superstrings has given rise to great activity in this field. But even back in the 1950s, physicists John Wheeler and Charles Misner were pushing ahead with a topological approach to space.

For many years Wheeler had been pursuing what he called

"Einstein's vision"—the belief that all of physics could be reduced to the twists and turns of space-time. Einstein had shown how the force of gravity could be replaced by the curvature of space-time. But if gravity is an effect of the geometry of space-time, then why not electricity and magnetism? Why not, indeed, matter itself? Einstein's vision was of replacing matter with hills and knots in the fabric of space-time and the forces of nature with regions of curvature.

This was the essence of Einstein's attempt at a unified field theory. But Wheeler realized that conventional non-Euclidean geometry was not rich enough to encompass Einstein's vision. There was simply not enough freedom to picture matter and all the forces of nature as the curvature of space-time. Wheeler and Misner therefore decided to break some of the rules of geometry and go to the deeper form of topology.

But would it be possible to create such things as elementary particles out of the properties of space-time alone? Think, for a moment, of a black hole. The black hole begins as a star whose mass and density is so high as to curve space-time tightly around itself. The curvature is so excessive that at the center, space-time becomes so distorted that its properties break down. In other words there is a tiny hole—what mathematicians call a singularity—at the center of every black hole. Indeed, to a mathematician a black hole is simply the result of a change in the topology of space-time, that is, the creation of a hole.

But suppose that you hover in your space ship just outside the event horizon (the point of no return) of the black hole. You experience an enormous gravitational pull but nothing else can be seen. There is no burning sun at the center of the hole—for no light can escape from its event horizon. It is not even possible to see a hole. All that any external observer is aware of is curved space-time and nothing else. Indeed the only physical properties that can be ascribed to the hole is that it has mass and, possibly, that it is spinning.

Now John Wheeler was struck by something curious. Here we have, he said, a massive spinning object that is nothing more than a hole in the fabric of space-time—a change from the normal topology. So a hole can behave just as if it were a massive spinning body. But would it be possible to create all particles that way? Is mass and spin nothing more than a certain property of holes in space-time? Wheeler called such objects "geons" and speculated that they could well be the fundamental particles of nature.

But particles like a proton and an electron have an electrical charge in addition to their spin and mass. Could electricity and magnetism also be described in terms of changing topologies of space-time? Think again of the black hole. Nothing can ever emerge from such a hole; anything that approaches too closely will fall in, never to reemerge. Some physicists began to speculate whether the exact reverse of a black hole was possible: a white hole which is forever radiating energy—that is, throwing things out. Wheeler and Misner had also begun to wonder what would happen if a black and a white hole could be connected directly—for example, if an extra "handle" could be used to connect one hole to another. This handle would not exist in our normal space but would be a way of connecting white holes to black holes. What one would see is a region of space-time to which matter and energy is attracted and, far away, another region out of which energy bubbles up. Energy that falls into a black hole would reemerge—possibly instantaneously—from a white hole that could be on the other side of the universe. Since the handle that connects the white hole to the black hole does not lie in our normal space-time the connection between white and black holes could well be instantaneous.

Wheeler applied this idea to the electron, conceiving it as a connecting hole in space-time. In conventional physics an electrical field is pictured as converging on an electron. But if it is a sort of hole, in the case of a negative electron the electrical field would seem to flow down into it. For a positive electron, the field would seem to flow out. Suppose, Misner and Wheeler said, we take this literally. Suppose that electrons are holes in the fabric of space-time through which the electrical field (or regions of curvature) can flow.

FIGURE 6–1

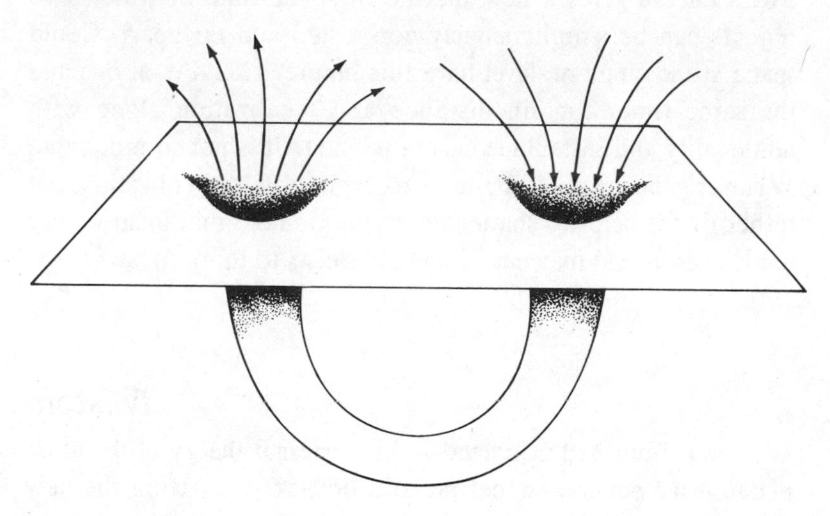

A wormhole in space-time. An electrical field enters the hole in one region and reappears at the other. The effect is identical to that of a negative and a positive charge.

These are not simply holes in space-time but wormholes. Just like a worm, the electrical field disappears down one hole (the negative end) only to pop up out of the other, positive charge. The fabric of space-time is threaded with these wormholes, each one representing a particular sort of elementary particle. Under this theory, it becomes clear why there are only two sorts of charges and why there are an equal number of positive and negative electrons—a wormhole has only two ends.

Wheeler's geometrodynamics, as he calls his topological theory, gives another view of nonlocality. From the perspective of conventional space-time, the ends of the wormholes can be a vast distance apart. Yet through the wormhole itself the two ends are in immediate contact; the connections between the ends are instantaneous. Travel by spaceship, and it may take years to go between wormholes, but fall into a hole, and you emerge instantly in another part of the universe.

While John Wheeler's vision of a quantum topology was never fully worked out, his ideas do remain fascinating. Indeed, Wheeler's picture of a universe threaded by wormholes like a

Swiss cheese gives a new picture of space-time in which two objects can be simultaneously connected and far apart. Could space at the quantum level have this nature? Could it, at one and the same time, contain distance and localization along with nonlocality and instantaneous connection? It is not so much that Wheeler's theory could be used to explain Bell's nonlocality, but rather that it helps to shatter the hypnotic hold that local reality holds over us and may encourage physicists to think in new ways.

Twistors

Wolfgang Pauli had criticized Bohr's original theory of the atom because it "put new wine into old bottles" by setting the new ideas of quantized energy within an older context of paths and orbits. Heisenberg's quantum theory broke with these classical ideas; nevertheless it continued to use old ideas of space and of time, clinging to a concept of space that had persisted since the time of Descartes and Newton. Quantum physics assumes, for example, that space is infinitely divisible, that it is continuous and can be broken down into smaller and smaller volumes. Indeed, the geometry of space is based upon the idea of the dimensionless point as its most fundamental object.

Geometry begins with the point. Building logically from the dimensionless point, it is possible to recreate all of geometry. Between any two points is a line. A line is defined between points, and it consists of an infinity of points. Take any two of these neighboring points, and yet another infinity of points lie between them. Infinities are nested within infinities, on and on in infinite regression.

But this idea of the continuity of space, with its foundation on dimensionless points, runs headlong into Heisenberg's uncertainty principle. In addition to position and velocity being noncommuting, the uncertainty principle also dictates that the more closely you pin down an object in space-time, the more uncertain does its energy become. In fact, if you try to pin things down to the dimension of a point, then the energy uncertainty becomes infinite. The very divisibility and continuity of space give rise to a bubbling, seething ocean of energy. The closer you examine space-time, the greater are the fluctuations in the energy it contains.

Einstein had taught that energy, like matter, causes space-time to distort and curve. At distances well below the sizes of the elementary particles, this fluctuating energy—a direct result of Heisenberg's uncertainty principle—causes space to fold and ripple. At even smaller distances, the very fabric of space-time begins to curl back on itself, creating bubbles and a veritable foamlike structure. What we take for continuity is an illusion brought about by the scale at which our elementary-particle experiments probe the structure of space. At shorter distances, physics must enter an entirely new region—at least according to current quantum theory.

There is, as yet, no true theory of this dynamic space-time of the subatomic world. No explanation exists for its dramatic new properties. It was this challenge that led Oxford mathematician Roger Penrose to develop his twistor theory of space-time. Penrose's idea was that quantum processes somehow generate their own space-time. Space is not a passive backdrop against which quantum events are played out, but is generated in a dynamic way. In other words, space-time and quantum particles both would emerge out of the one unified theory.

What should be the starting point of this new quantum geometry? Penrose firmly rejected points as the foundations of his geometry and selected instead a generalization of the line called a twistor. Twistors are lines of infinite extent, in some ways like twisting light rays. These extended twistors become the most elementary objects in Penrose's twistor geometry and, in contrast to the way in which lines are built out of points in conventional geometry, points are built out of the conjunction of twistor lines. At its most fundamental level, twistor space is nonlocal. Indeed, the idea of locality becomes more complex and must be built out of logically more primitive nonlocal elements.

Penrose's twistor approach is still in the developmental phase, but already it is becoming clear how our space-time could be built in a very complex way out of an underlying twistor space. Indeed, Penrose's theory may be a way of unifying relativity and quantum theory, for the space-time that emerges is curved, and this curvature is naturally related to the presence of matter—quantum particles. The space-time we are more familiar with will emerge as a limit out of the underlying twistor space. In fact, our space-time may no longer be something simple but could be quite complex, like a multilayered onion, displaying both locality and nonlocality.

SPACE AS PATCHES OR OVERLAYS?

Penrose's twistors are a dramatically different way of portraying the structure of space-time. Will twistors provide the clue to the nonlocality inherent in Bell's correlations? In relativity, space-time is compared to a patchwork quilt in which the twists and curves of space-time are created by joining together a series of tiny, regular patches. Within each patch, things are well behaved, but at the edges, the "patterns" of neighboring patches do not match up. This series of discontinuities gives rise to the curvature of space-time and the phenomenon of gravity.

This is a purely local approach to space-time, in which the bigger, curved picture is built out of a series of smaller, local elements. Is there an alternative picture that would display nonlocality? Rather than a patchwork quilt, space-time could be compared to a color photograph in a book. The color printed on the page is based on a series of "color separations." This means that the original photograph is broken up into four images, each printed in either magenta, yellow, cyan, or black. When superimposed, these four images give the impression of perfect color. The actual picture is generated through four separate acts of printing, each of which uses one of the four colors of ink. Thus, each color separation contains different information, but the total effect is created by combining them.

It may be possible to picture space in an analogous way, not as a combination of local patches, but as a superimposition and overlapping of more primitive spaces that, taken together, produce the overall effect of space-time.

Just as the atom was once thought to be simple but has been revealed to have a complex internal structure, just as the elementary particles were once believed to be fundamental but are now thought to be composed of even simpler objects (superstrings), so too space-time may be built in a complex way out of more primitive elements. Some of these elements may be, as Penrose suggests, nonlocal. Distant systems that are directly connected through the nonlocal layer of space will then display correlations that are inexplicable from the viewpoint of local space.

A Nonlocal Space

The great elementary-particle experiments of the last decade, involving bigger and bigger particle accelerators, have all concentrated on probing smaller and smaller spatial regions. But what about the possibility that space could have some new and unforeseen long-range properties? The current experiments of physics would never reveal this novel property, for they are all based on probing quantum effects within smaller and smaller regions. However, elementary-particle physicist Tsung-Dao Lee has suggested that experimentalists should look at the effect of distributing high energy, or matter, over an extended area. Lee calls this "vacuum engineering," using the elementary-particle physicist's term *vacuum* for empty space-time. "If indeed we are able to alter the vacuum," he writes, "then we may encounter some new phenomena, totally unexpected."*

Imagine quantum experiments of the future that are designed to peel away the onion skins of space. Bohr had pointed out how different experimental arrangements will reveal different aspects of a quantum system. One arrangement gives information about

*T.-D. Lee, *Particle Physics and Introduction to Field Theory* (New York: Harwood Academic Publishers, 1981).

position; another, incompatible with the first, gives information about the noncommuting parameter of velocity. Could there be other forms of experimental arrangement that are designed to reveal different, even mutually incompatible, aspects of space? Like a holographic plate, space-time may contain nonlocality (information about the whole scene) within each local part. Space may be created out of the overlapping and superimposition of a series of "color separations," each with its own more primitive properties. Different experiments may reveal different aspects, different levels of this superimposition of properties. Where a twistor space is involved, the fundamental levels will be essentially nonlocal. Just as elementary particles exhibit both wave and particle natures, so too space-time may contain both local and nonlocal aspects.

Penrose has argued that space-time is highly complex and created out of subquantum processes. In the double-slit experiment, for example, does the electron really split itself in two to go through both slits at once? Or within its own personal space, is everything perfectly regular but with only a single slit appearing in the electron's path? Rather than the electron suffering a schizophrenia, the experimental apparatus could be, in the electron's opinion, distorted and transformed. According to one perspective, the electron is in two places at once; according to the other, the apparatus transforms to present only a single slit. Space-time would represent a superimposition of these different views, so that a full understanding of the meaning of "quantum reality" would depend upon discovering the complex relationships between space-time and matter.

Space-Time and Quantum Actuality

Penrose maintains that space-time plays a key role in that most fundamental of quantum events—the mysterious process whereby many quantum potentialities become a single actuality. Quantum theory deals in probabilistic outcomes, predicting relative probabilities for a variety of different possibilities. But how are these potential events translated into actual happenings, into the click of a Geiger counter?

The previous chapter explored a number of alternative quantum interpretations in which this collapse of potentialities was variously ascribed to multiple universes, interactions at the large scale, or even human consciousness. Penrose has quite a different idea. Quantum events must all occur in space, but this space, he suggests, is actually created at the quantum level, possibly involving events in twistor space. At this subatomic level, a number of possible quantum spaces coexist, all superposed.

These spaces could be thought of as the germs out of which our own large-scale space is built. At scales far below the atoms, these various space-germs coexist on an equal footing; however, as their size increases they reach a point where a concrete space actually crystallizes out. This has the effect of collapsing potentialities into a single actuality. This critical point occurs when a single graviton—a single quantum of space-time curvature, a single quantum of gravity—forms. An actual quantum space-time element would expand out of the superimpositions of pre-space, and, as with the expansion of a bubble in a bottle of Perrier, there would be no going back—the quantum event would have been actualized.

Something analogous may occur in the Bell experiment, says Penrose. A combined measurement of both photons, at *A*, and at *B*, is necessary to effect the creation of a space-time structure that connects the two quantum events. At the start of the experiment, space-time will be only partially formed. It is like a developing negative in a darkroom bath, except that, at this stage, the image may develop in several possible ways. Finally, when the measurements are completed, this nonlocal space fixes itself. According to Penrose's picture, physicists can expect to find novel properties for this pre-space; only in the large scale of many quantum particles and events would these overlapping and superimposing views condense into our everyday space-time.

Nonlocality and Unbroken Wholeness

Nonlocality may be a bizarre new idea, yet it has been possible to invent a series of images to illustrate what it may mean. Roger Penrose has even developed the mathematics of a quantum space that is based on something fundamentally nonlocal—a twistor. Does this nonlocality actually operate at the quantum level so that the two photons in Alain Aspect's experiment, although far apart from the perspective of a scientist in the laboratory, are at another level connected? Such nonlocal connections could, in fact, stretch throughout the entire universe. They would give yet another meaning to quantum reality, another possible answer to Einstein's question about the moon. In one sense the moon is distant; in another it is in constant connection with us. The moon exists, and we are always in contact with it, always observing it. Yet thanks to quantum theory and Bell's theorem, our understanding of the meaning of existence and reality must change dramatically.

THE QUANTUM POTENTIAL

Bell's theorem has spelled the end to hidden mechanical variables and local ideas of reality. Either the idea of reality must be abandoned altogether, or we must learn to think in new, nonlocal ways. One possibility along this latter line has been proposed by David Bohm. It makes a total break with conventional quantum theory. It is based upon a concrete reality and complete determinism, yet at a fundamental level is nonlocal in nature. Most physicists would not accept Bohm's quantum potential, however, it does give us powerful insights into the meaning of nonlocality.

David Bohm is one of those handful of physicists who have never accepted the quantum theory. Born in the generation that followed Einstein, Bohr, and Heisenberg, Bohm studied with American physicist J. Robert Oppenheimer but felt uneasy with the Copenhagen interpretation. Bohm wrote *Quantum Theory*, which became a classic among physicists, in an attempt to come to terms with the orthodox theory. Nevertheless, he was unwilling to go along with quantum theory's wholehearted rejection of causality and the reality of the electron. In the end, it was his discussions with Einstein, held at Princeton in the early 1950s, that persuaded Bohm to devote his energies to an alternative approach. Over the following decades, Bohm became one of the most thoughtful and far-reaching critics of the orthodox interpretation. Other physicists may have argued against Bohr's interpretation, but few understood their opponent's position as deeply as David Bohm.

In addition to his philosophical writings, Bohm had been working on his own mathematical version of a quantum theory that provides a deep insight into the meaning of nonlocality and gives a new way of looking at Bell's result. Bohm's theory comes in two levels or versions, the second being more speculative and, in a sense, deeper than the first. This second version views nature in terms of a constant flux of fields that enfold and unfold to produce the various phenomena of physics. In his earlier version, which is more completely developed and discussed here, the electron is viewed as a purely realistic particle that, like any other particle in large-scale physics, has a definite velocity and position. Knowing that velocity and position at each instant of time makes it possible to calculate the path of the electron. In contrast to the Copenhagen interpretation, this theory holds that the electron is real; it has a path and possesses properties even for those aspects of its reality that Heisenberg claimed were noncommuting. In Bohm's theory, there is no ambiguity about the electron or, for that matter, about any other quantum system.

As the electron moves along its path, it is pushed and pulled by electrical and magnetic forces, just as any other real particle would be. So far, the theory is no different than any other "classical" theory of nature and could certainly never account for those curious quantum features like Heisenberg's uncertainty, the double-slit experiment, and the Bell correlations. Some entirely new feature is needed.

This new feature comes from Schrödinger's equation. Schrödinger originally believed that his wave equation could save quantum theory from Heisenberg's abstraction by treating the electron as a real wave. In Bohm's opinion, Schrödinger was correct about the reality of the electron but not about its wave nature. The electron, says Bohm, is a particle, not a wave, but its motion is determined in a curious way. Bohm was able to show that Schrödinger's equation can be written in an entirely new way. When it comes to numerical results, the old and new forms of the equation are entirely equivalent; it is their meaning that has changed.

Bohm showed that, in its new form, the Schrödinger equation describes a real electron that is pushed and pulled by the usual electrical and magnetic fields. In addition, an entirely new sort of force has been introduced. This force is called the quantum potential. This quantum potential turns out to be responsible for all the curious new features of quantum theory.

Bohm's quantum potential is unlike any potential or force that has been discovered in physics to date. Gravity, magnetism, electricity, and the nuclear forces are all local forces, and all act in a mechanical way by pushing or pulling. Because of the earth's gravitational pull, the moon does not fly off into space. A compass needle points north because of the magnetic pull of the earth, and the Japanese maglev train hovers above its rails due to magnetic repulsion. In all cases, the stronger the force field, the bigger its push or pull. And, since the strength of a force field falls off quickly with distance, the farther you are from the source of a push or pull, the weaker is its effect.

These are the forces that physicists have worked with for more than two hundred years. In fact, our whole picture of nature is based on the behavior of these forces. Because the effects of pushes and pulls fall off so quickly with distance, the world can be divided into separate objects.

It is the nature of these pushes and pulls that gives the world its mechanical aspect and makes the idea of locality so commonplace. When things are close together, their mutual influence is large, but pull them apart, and they become virtually independent of each other. This is essentially the meaning of locality: effects that are strong within a given region of space fall off outside, so that it makes sense to divide the world into separate, self-contained systems that interact by forces and signals that fall off rapidly with distance.

Into this comfortable picture, David Bohm introduced a totally new idea—the quantum potential. This quantum potential is unique because it does not work in a mechanical way by pushing or pulling but has more the nature of a guide wave. Furthermore, its effect does not fall off with distance. Bohm has given the analogy of an ocean liner. The great power of its engines pushes the liner through the ocean. But the actual direction taken by the liner originates in something far weaker and more subtle—the radar signals that indicate oncoming ocean traffic and the centers of storms. It is this weak energy of the radar signal that provides the information that actually guides the ship by directing and focusing the much greater power of its engines. One could even imagine a computer control in which information from the radar signal acted to steer the boat directly.

In a similar way, the quantum potential directs the electron along a path. The broader outlines of that path are determined by the mechanical pushes and pulls of magnetic and electrical fields, but the actual details are determined by the quantum potential. Because of the guiding nature of the quantum potential, its effects do not depend upon its strength. Therefore, the effect of the quantum potential does not fall off with distance and can be important at great separation. Already, the connection with Bell's theorem begins to come into focus.

An important feature of the quantum potential is its extreme sensitivity and complexity. Since the quantum potential does not fall off with distance, it is responsive to all sorts of changes and effects happening around it. The electron's path is totally deterministic, since it is governed by the quantum potential, but because of the sensitivity of this potential, the actual path turns out to be so complex as to be unpredictable. The electron's motion, guided by the quantum potential, is an unpredictable "jitterbug." Although it always has a definite position and velocity at any one time, these values are constantly changing. For this reason, the future of the deterministic quantum system is indeterminate!

Attempt to measure the position of the electron, and the act of measurement itself produces an uncontrollable change in the quantum potential. As Heisenberg had first suggested in his famous microscope experiment, the act of measurement disturbs the quantum system and changes its values. If you are acquainted with the ideas of chaos theory through books like Gleick's *Chaos* and Briggs and Peat's *The Turbulent Mirror*, you will recognize that the electron's path has much in common with a chaotic system—complexity and inherent unpredictability of its movement, and extreme sensitivity to external effects. Just as in conventional chaos theory, indeterminism, unpredictability, and uncertainty have their origins in causal, though extremely sensitive and complex, systems so the novel features of quantum theory can be explained in terms of the complicated and sensitive quantum potential.

The famous double-slit experiment also can be understood in terms of Bohm's causal interpretation. Because the quantum potential does not fall off with distance, the path of an electron or photon as it leaves its source is determined by the whole of its environment, in particular by the two distant slits. Using a computer, it is possible to calculate the possible paths allowed by the quantum potential.

The following diagram shows a complicated series of paths spreading out from the source and going through one or the other of the slits. Starting from a particular point, the electron will go through slit *A*. Another path begins an infinitesimal distance from the first and this time takes the electron through slit *B*. The fate of each electron is as if decided by a coin toss for, depending on its exact velocity and position as it leaves the source, it will move along one or another of these paths. An infinitesimal difference in its velocity, and it will find itself moving through a different slit.

FIGURE 6–2

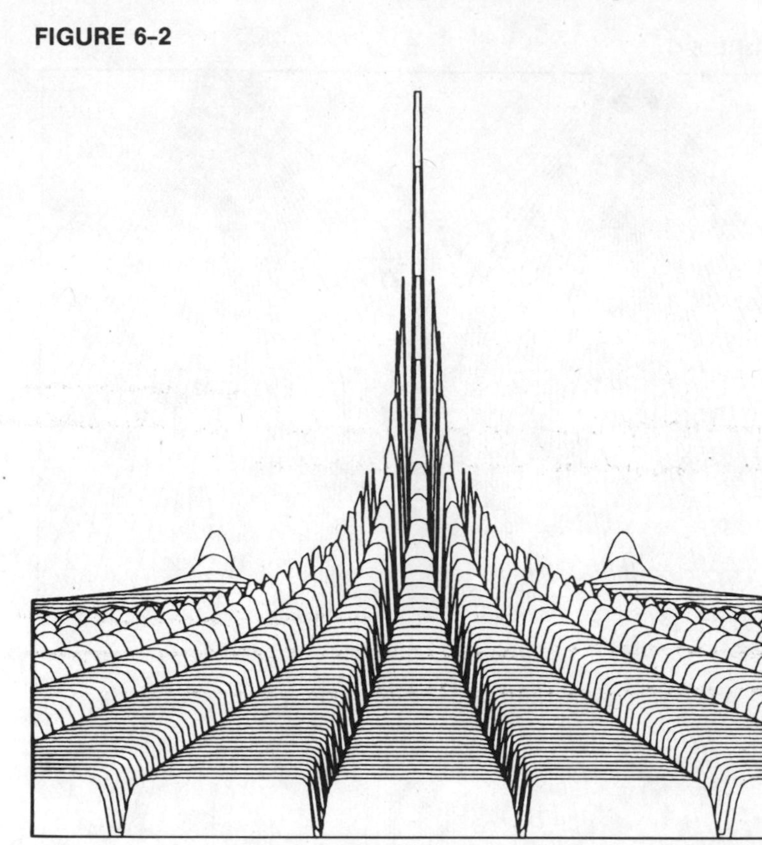

The quantum potential corresponding to the double-slit experiment is determined by the whole apparatus.

One electron after another is shot toward the slits. The path in each case is real and deterministic, yet because of the complex nature of the quantum potential, it is impossible to predict through which slit the electron will travel. But notice that the various paths have a periodically bunched character as they near the screen. Over many experiments, the electrons tend to hit the screen in this periodic bunched manner so as to give the appearance of an interference pattern. In Bohm's view, the electron does not have a wave nature; rather, the particular properties of the quantum potential simulate the appearance of wave interference.

A significant feature of Bohm's interpretation of the double-slit experiment lies in its wholeness. Niels Bohr had constantly

FIGURE 6–3

The various trajectories that can be taken by an electron after it has passed through the double slits. Note how these trajectories have a periodic bunching which produces a pattern on the screen that has the appearance of wave interference.

stressed quantum wholeness and the holistic nature of any quantum measurement. Now the underlying reason for this wholeness is revealed. The quantum potential, which is ultimately responsible for the patterns on the screen, is a feature of the whole of the apparatus. Cover one of the gaps in the screen, and the quantum potential changes—the interference pattern vanishes.

In refuting the EPR paradox, Bohr argued that the disposition of experimental apparatus determines what can be measured. To measure the second of a pair of noncommuting properties, it is necessary to change the apparatus in the laboratory, and, in Bohr's words, that has "an influence on the very conditions which define the possible types of predictions regarding the future behavior of the system."

In terms of Bohm's causal interpretation, this begins to make sense. Any quantum measurement is an unanalyzable whole. It is unanalyzable because the quantum potential that guides the motion of an electron is produced by the whole experimental arrangement and includes its most distant parts. Since the effect of this potential does not fall off with distance, it is impossible to break the apparatus and the quantum system into separate parts. Clearly, any change in the arrangement of the apparatus is going to alter the quantum potential in an unpredictable way. Bohm therefore gives a mathematical justification for Bohr's intuitive idea of quantum wholeness and shows that any change in the apparatus will totally and unpredictably alter the context in which a particular measurement is defined and thereby prevent a unique prediction of the system's future.

Quantum wholeness also features in Bell's theorem through the quantum correlation that exists between distant systems. In Bohm's picture, this correlation can be given a realistic and causal interpretation. Clearly, whatever is measured in region *A* is determined by the quantum potential, and the form of this potential is in turn determined by the whole experimental arrangement. Since the effect of the potential does not fall off with distance, any measurement made at *B* must indeed have an effect at *A*.

This brings us back to Einstein and Bohr's earlier debate. In setting up their paradox, Einstein and his colleagues had ruled out by reason of distance any possibility of *B* influencing *A*. Bohr argued, "In the measurements under consideration any direct mechanical interaction of the system and the measuring agencies is excluded." In his choice of wording, Bohr could almost have had the quantum potential in mind, for all mechanical interaction, all pushes and pulls, all conventional fields of force are eliminated by the separation of the two systems. What is not excluded, however, is what could be called a "nonmechanical influence"—the effect of the quantum potential.

The additional nonclassical correlations first discovered by John Bell are, in Bohm's opinion, a direct manifestation of the quantum potential and suggest a new way of looking at nonlocality. Since the effects of the quantum potential do not fall off with distance, they are unlike anything that has previously existed in physics; in fact, they defy normal physical intuition. They are truly nonlocal, since they are not diminished in any way by distance or separation, as is the force of gravity or magnetism. The quantum potential becomes a way of linking or correlating distant events, not through some force of pushes or pulls, but through a more subtle form of "guidance."

But if what happens at *B* has an instantaneous influence on *A*, a great distance away, then isn't the potential acting as a sort of signal?

For the quantum potential to work as a true signal, and generate Einstein's causal paradoxes, one must be able to modulate the quantum potential in such a way that it can transmit useful information at will—for example, to switch the potential off and on like Morse code or to modify its signal in some way as with the carrier signal of a radio broadcast. If this could be done in such a way as to change the trajectory of an electron, then, indeed, superluminal connection would be possible. But it is easy to show that any influence on the form of the quantum potential has a totally unpredictable effect. The potential is so sensitive that any attempt to use it as a signal would result in a totally garbled message. While Bohm's causal interpretation is realistic, its quantum potential can never be used as an intelligible signal. The quantum potential has nothing to do with faster-than-light signals but implies a totally nonlocal reality.

The idea of a potential that works by guiding an electron rather than through mechanical pushes and pulls appears unfamiliar at first sight. Bohm, however, has written about the quantum potential in terms of information. The quantum potential gives form to the raw energy that moves the electron. Bohm sees this process of a subtle energy giving form to a more powerful one as universal.

Bohm calls this giving of form "in-form-ation." Indeed, it is similar in many ways to the usual meaning of the term *information*. After all, the quantum potential is the sum of all the particularities of the arrangement of experimental apparatus in a laboratory. In this sense, it carries that information within itself and uses it to give form to the motion of the electron. For this reason, Bohm refers to the action of the quantum potential as it determines the movement of an electron through a slit as "active information." For Bohm, active information has something in common with the activity of mind and suggests that the universe could be thought of as an organism of great subtlety.

To many physicists, Bohm's causal interpretation is unacceptable (although it must be admitted that most physicists are not too clear about what Bohm's theory actually says). Over the decades, physicists have grown so accustomed to wave-particle duality and quantum uncertainty that it looks like a backward step to propose a realistic electron acted on in a deterministic way. Many physicists doubt that the universe could really work in this way and find the causal interpretation unconvincing.

There is no denying, however, that the theory provides valuable insights through its mathematical formulation of quantum wholeness and nonlocal interconnectedness. These are the ideas that Niels Bohr talked about but could never demonstrate in any formal way. Thanks to the quantum potential, it has become possible to give an actual picture of nonlocality and quantum wholeness.

Bohm's more recent formulation of the causal interpretation goes beyond the idea of an electron as a material body and sees it in terms of the constant enfolding and unfolding of a field whose form is determined by a super quantum potential. Again the universe is pictured as a nonlocal reality. It is nonlocal because what occurs in any region of space is determined by the form of the entire universe. Nature is a single organism that has its total reality enfolded within each tiny region of space.

A CONNECTED UNIVERSE

Bohr's quantum wholeness, Bohm's quantum potential, and the idea of nonlocality that can be inferred from Bell's theorem are all new ways of looking at the universe. They suggest, at the atomic level at least, a universe with a remarkable degree of interconnectedness. But can this unbroken quantum interconnection justifiably be projected onto the entire universe? Some thinkers believe that it can. Indeed, nature was generally seen in this unified way until a more mechanistic science came along in the seventeenth century.

Newtonian science, which pictured nature in terms of independent bodies connected by forces, proved remarkably powerful and produced deep insights for over two centuries. But is this the only way to look at nature? Is it possible to discover a role for something that is nonlocal and holistic? The problem with an exclusively holistic approach is to account for the fact that the world appears, at one level at least, to be composed of separate, independent objects. However, the emerging sciences of nonlinear systems suggest new ways of reconciling holism and separateness within a single, unified vision.

The whirlpool or vortex in a river, for example, has a definite location in space and persists in time. It is even possible to generate in the river special sorts of waves, called solitons, that behave in many respects like particles—even to the extent of colliding with each other. Yet these vortices and solitons have no independent existence apart from the river that supports them. The vortex exists through the act of being constantly created. At every instant, it dies and is reborn. The fountain in a city park and every living cell in the body are not fixed, rigid structures but maintain their features through constant flux. The flow of water through the fountain or whirlpool gives it permanence; the flow of matter through a cell keeps it alive.

Such systems are essentially wholes, yet out of their overall movement and change they give rise to quasi-independent parts. These parts are, however, constantly being generated by the system itself. They are like the individual aspects of a landscape that are nevertheless unfolded out of the whole of a holograph. Each part of the scene emerges out of the whole holograph, yet the whole is contained within every part.

A similar vision extends across the physical universe. It suggests that the universe is not a dead thing, composed of separate parts in mechanical interaction; rather, it is organic—an almost living thing in which each aspect subsists on some underlying activity. At one level, we perceive independence and material objects acted on by pushes and pulls; at another, we see flux and constant change.

David Bohm has introduced the ideas of the implicate and explicate orders. The explicate order refers to the surface of things, to a mechanical world of pushes and pulls. The implicate order, by contrast, involves an enfolded reality. The implicate order lies beyond the categories of space and time and is, Bohm believes, a more appropriate way of ordering the quantum theory. In a sense, while Newtonian physics describes the explicate world, quantum theory is science's first attempt to come to terms with the implicate. It is at the implicate level that Einstein's question about the reality of the world must be answered.

At present the quantum theory is limited by concepts, mathematical forms, and a language that are all geared to discussing the explicate world. To achieve a proper understanding of quantum theory, it may be necessary to move into an implicate level of description. But although scientists like David Bohm and Roger Penrose are developing ways of discussing nonlocality, quantum connectedness, and the implicate order, the basic approach is still foreign to many scientists.

Nevertheless, even if mainstream scientists continue to think in traditional ways, there are indications that some physicists are beginning to explore new ways of understanding the universe.

Indeed, people have always had an intuitive sense of their interconnectedness to nature. The Native American experiences *skanagoah* or great peace when alone in the woods, an electrifying awareness that involves a sense of unity with all of nature. Similar experiences are reported by artists and mystics of all cultures. In fact, it appears more natural to view the universe as connected and immanent than as mechanical and separate. The philosopher Edmund Husserl argued that the crisis facing modern men and women lies in the meaninglessness of the world around them. He traced the root of this problem to the Cartesian-Newtonian desire to objectify nature. But when nature becomes an object, then human values and relationships are sacrificed. The result is an empty, meaningless universe.

Our direct experience tells us that we are in some way connected to the cosmos, but until recently, science indicated that such connections had no basis outside the normal interactions of physics. Working with Wolfgang Pauli, psychologist Carl Jung developed his own idea of an "acausal connecting principle" called synchronicity. There are patterns in nature, and connecting patterns in consciousness, Jung claimed, that are not generated by any mechanical cause. Moreover, these patterns often have a numinous meaning for us. If the universe is spanned by meaningful patterns, this suggests an activity of meaning within nature.

It is unfortunate that the notion of synchronicity has today become mainly associated with bizarre coincidences, for Jung was serious about proposing a new principle in nature. The idea of connections or correlations that lie beyond causality resonates with Bell's own powerful insight. Just as Einstein added time to space to produce the much deeper concept of space-time, so Jung proposed completing causality by adding a noncausal link. Certain patterns in nature, he argued, are linked in nonmechanical ways to form a "causeless order." Bohm, for his part, later proposed that active in-form-ation actually works within nature and in so doing extends the material universe into subtle areas that begin to blend with mind.

Jung's synchronicity extends beyond the material, for its

patterns are meaningful and are echoed in both mind and matter; he also termed his acausal connection a "meaningful orderedness." Thanks to Pauli's connections, Jung's ideas were brought to the attention of some of the leading physicists of the time, including Heisenberg, who certainly did not dismiss them out of hand.

Bell's notion of an acausal correlation, a connection between distant events that does not involve any mechanical signal or interaction, implies a new nonlocal view of space. While Bell's arguments are confined to the atomic scale, this does not mean that a general nonlocality may not be important in the large scale as well. Jung himself may have had some inkling of such a vision when he wrote, "Knowledge finds itself in a space-time continuum in which space is no longer space, and time time."

In suggesting that nonlocal connections may have a universal application, I am not naively proposing an extension of the Bell's theorem correlations; that would be an unjustified extension of Bell's result. Rather, I am suggesting that they may be a particular manifestation of something that is far more general. Indeed, nonlocality may be a better way of describing the world than exclusively in terms of local, causal connections.

In a Newtonian world, systems are built up out of interacting parts, like a model built with a Lego set. But the whole approach of the quantum world is to speak in terms of an overall form. Schrödinger showed that a quantum system can be described in terms of a wave function; moreover, it is the overall *form* of this wave function that determines the behavior of the quantum system. Wolfgang Pauli, for example, showed that the form of the wave function must have a particular symmetry, and, in a sense, symmetry requirements also lie behind the Bell correlations. Nature, at the quantum level, deals in global forms and patterns, in a sort of information, moreover these forms cannot be defined locally but are essentially global.

Some sort of nonlocality may turn out to be a useful way of viewing the structure of molecules and crystals, even the structure of space and time. (Roger Penrose has given some hints as to why this may be true.) Our own bodies also can be seen in an interconnected way. Traditionally medicine divides the body into organs and causal interactions. Yet research on the immune system suggests another sort of description involving a common meaning that floods the body and makes for health. Practitioners of holistic medicine believe that when this meaning becomes confused or blocked, illness will result. Certainly the causative model of sickness is at one level appropriate, for different organisms do invade the body and cause specific diseases. Yet even though our bodies are constantly subject to such attacks, we do not always become ill. When we "have" an illness, it seems to involve a general breakdown within the immune system. To put it another way, the meaningful order of the whole body loses its coherence and allows sickness to develop.

A healthy body is one suffused with an active meaning, and it is most appropriate to describe this in a nonlocal way. Indeed, the whole notion of local, Cartesian space may be meaningless when applied to the body. The body—and indeed each individual cell—creates its own rich space, together with its own rhythms. The body is a system of connections, correlations, flows, interactions, and correspondences that, in good health, work in harmony and are not blocked. The body's description is multilayered, both local and nonlocal, causal and holistically correlated.

The same sort of description applies to society, in which the activities of the individual flow out of the general values, meaning, and significance of society as a whole. At the same time, society is itself created from the values and meanings of each individual member. The society and the individual are like the egg and flour in a cake; each is enfolded into the other, each contained within the other.

Events and processes become interrelated in entirely new ways, and a new set of images is needed to describe this more holistic vision. A useful analogy for this way of seeing is to contrast it with a stone thrown into the middle of a lake. Near the center of the lake, large ripples are produced, which spread out and dissipate. At the edge of the pond, these ripples are so small that they become lost in the random disturbances of the water. In terms of conventional physics, the order of the original splash has been transformed into the general chaos of tiny ripples, disorder (entropy) has increased, and the arrow of time has marched from past to future.

But it is theoretically possible for things to move in the opposite direction. Suppose that nonlocal correlations are introduced. Then very small ripples all around the edge of the pond will become correlated in a nonlocal way. Rather than beginning with a violent, local event, something more gentle yet global forms the starting point. Because of their correlations across the entire edge of the pond, these small ripples will now interfere in a constructive fashion, which means that their effect begins to accumulate as the ripples move inward. The effect is of tiny waves that flow inward from the edge of the pool, growing until they reach the center and produce a large splash.

The whole thing looks like a film of a normal splash but played in reverse. The arrow of time appears to go backward as ripples spread inward rather than outward. Disorder and chaos appear to have given way to order. At present, this example is simply a thought experiment, yet the curious phenomenon is perfectly possible once nonlocal correlations are permitted. In fact, something analogous may be going on in many aspects of the natural world, for the high degree of information present in the nonlocal correlations of all the ripples at the edge of the pond allows for this seeming reversal of the law of entropy. To borrow David Bohm's term, active information transforms apparent chaos into order. Thanks to nonlocal correlations, it becomes possible for a very gentle action distributed over the whole system to focus into a particular part.

A similar analogy can be applied to the human body. A healthy person experiences an overall flow of meaning, partly conscious, partly unconscious, throughout his or her life. This meaning has a powerful activity throughout the body and could be compared to the nonlocal correlation of all its parts. Meaning constantly acts to produce tiny, gentle changes throughout the body, so that in the case of injury, the person is able to focus healing ability within a given part.

There are also hints that the brain may work through global correlations of its activity. Decades of research into the nature of memory indicate that it is not localized in one particular region of the brain but appears to be distributed over its whole system. Understanding memory, and other activities of the brain, demands a nonlocal theory. Neuroscientist Karl Pribram proposed, for example, that memory is distributed within the brain in much the same way as an image is enfolded across a holograph. Damage to a normal photograph results in the loss of part of the image, but in the cases of a holograph, the totality of the entire image is preserved, although some degree of definition is lost. In the brain, Pribram argued, injuries in localized areas, or even peppered across the brain, do not result in the loss of selective memories. For example, destroying one part of the brain does not result in the loss of memories for a particular period of a person's life.

While Karl Pribram's idea has yet to be accepted by mainstream neuroscientists, it is certainly suggestive and has sparked other theories along similar lines. Another way of thinking about a delocalized memory is in terms of what are called neural nets. These are the complex interconnections of nerve cells that are partially responsible for the brain's activity. (The brain's functioning is also governed by a set of neurochemicals, including the neurotransmitters, and a true theory of the brain must take both its electrical and chemical activities into account.) In a neural net, each nerve ending is interconnected to several thousand other nerves. New information changes the sensitivity of these interconnections, which in turn leads to information being processed in different ways. So the brain is never fixed and, with each new experience, the fine details of its structure change.

Let us speculate what would happen if some form of nonlocal correlations also existed within the brain. Memories could then be stored nonlocally as very tiny but globally correlated changes in sensitivity. In this way, an electrical activity would sweep inward, like the inward moving ripple in the lake, to concentrate in a particular region. It would spread outward again and then concentrate within some other region. The brain's activity would correspond to an activity of unfolding and enfolding based on subtle nonlocal correlations. Since neural nets can be simulated on a computer, it would be interesting to perform computer experiments that allow for an artificial form of nonlocal correlation between elements of the net.

Acausal connections and nonlocal correlations may be a powerful new way of understanding the universe. While the correlations of Bell's theorem are specific to quantum systems, they may be manifestations of something even more general. Space itself may be built up in ways that lie beyond simple locality. The ideas of nonlocality may also be necessary for a proper understanding of the operation of the brain, body, ecology, and even of society itself.

Indeed, a new form of activity, which I have called "gentle action," could form the basis of a more appropriate response to today's challenges in economy, ecology, society, and international relations. The environment, for example, is composed of highly complex ecological systems that interact in nonlinear ways. Such systems have a wide variety of behaviors that range from stability and rigidity to change to sudden bifurcation points and extreme sensitivity. The very complexity of these systems may make them impossible to model in any satisfactory way. In addition, any human intervention often leads to unpredictable results.

The traditional response to social, economic, and environmental threat is to analyze the situation, isolate the problem, propose a solution, and then implement it in an active way. But such solutions are generally confined to a particular region or, in a fragmentary way, to some particular aspect of the system. Experience suggests that such "solutions" can have unexpected effects and that, when it comes to highly complex and interrelated systems, the cure may even be worse than the problem itself!

An alternative approach is to operate throughout the whole system in a gentle, nonlocal way. Rather than attempting to change the direction of a process or actively oppose some particular effect, the key would be to operate in a subtle but global way and seek to restore harmony through gentle correlations. In the case of the lake, a large effect in the center can be produced not only by directing a large amount of energy into this region but also by working at the edges and applying a gentle form of correlation to all the ripples. The "solution" is not generated in a fragmentary fashion by focusing on a single region but arises naturally out of the activity of the system as a whole. The key to the serious problems that face us today may lie not in a call for immediate action but in careful and sensitive observation and a gentle instinct for balance and harmony.

CONCLUSION

The quest of this book has been for the nature of quantum reality. It has become clear that, thanks to quantum theory and Bell's theorem, physics has finally broken with a reality based upon local and mechanistic concepts. Quantum reality, if it exists, must incorporate ideas of nonlocality, wholeness, and enfoldedness. Indeed, what seems to be called for is the development of new mathematical forms and a more subtle use of language to investigate this new order in nature.

But physicists are not the only thinkers to have investigated the nature of reality in this century. For the last fourteen years of his life, while writing *A la recherche du temps perdu* (*Remembrance of Things Past*), Marcel Proust ceaselessly struggled with the meaning of reality. Proust investigated the way memories and perceptions surfaced in his mind, the way he created the world around him every morning as he surfaced from sleep, what gives things their reality, the nature of time and memory. Again and again, the answers he uncovered are expressed in terms of overlaying, overlapping orders and the superimposition of different experiences, memories, tastes, and smells. Proust's memories and experiences do not so much add up to a single tangible reality but to realities that coexist, often in paradoxical ways, as when he remembers the different cities of Italy that he has visited both in reality and imagination. These real and imagined cities coexist in the mind.

In an analogous way, the actual buildings that have been created on Abraham's Rock in Jerusalem coexist in space and time. These places of worship—Jewish, Christian, and Islamic—are recognized by religious people as the one holy place. They were built at different historical periods, yet they coexist in the present moment, each enfolded within the other, each equally real.

So in a real Jerusalem and in the pages of a French novel, place and reality, space and existence overlap and merge. In reading Proust's great book, one is struck by the fact that the same task faces physics today—that of making sense of a multilayered reality in which are entwined the complex levels of space and time. Has the task of modern physics therefore become one of deconstruction? Does it involve stripping away layers of illusion created by outmoded theories, language, and paradigms and reaching toward some deeper truth?

We began by asking whether the electron is real. Does it only exist when we are observing it? Does it have properties independent of us? Now we realize that in asking these very questions, we were restricted to a particular mental mold and influenced by a particular view of reality, things, properties, causality, space, and time. We were mesmerized by a model of reality involving well-defined objects that is two centuries old, by a mathematics devoted to explicate relations, and by a language that evolved to serve in our large-scale world. Before quantum theory came on the scene, physics was concerned with a reality that could be touched, sensed, and tasted. It relied on words like *cause* and *effect*, *near* and *far*, *distance* and *proximity*. Today all this is changing and making way for a more subtle reality and a space that is multilayered.

So is Einstein's moon really there when no one looks at it? In one sense, what we call the moon is an aspect of a reality we have created though our physical senses and scale of being in the world. Quantum theory has taught us that whenever we do an experiment, we create a context that determines which sort of properties can be measured—wave or particle, position or velocity. In addition, our particular theories and indeed the language we speak condition the interpretation of these experiments. In this sense, we create the reality we see.

On the other hand, we are not free to observe any properties we desire. The quantum world has its own order and laws that are independent of us. The quantum world involves an order that we have partially created yet are still attempting to understand. Its constantly unfolding order is independent of our wishes and desires; it is not created by our consciousness or even our acts of observation. In this sense, quantum reality is both independent of us and dependent on the contexts we create.

Einstein's moon exists. It is linked to us through nonlocal correlations but does not depend upon us for its actual being in the world. On the other hand, what we call the moon's reality, or the electron's existence, depends to some extent upon the contexts we create in thought, theories, language, and experiments.

Quantum theory has opened a new door onto reality, a totally different way of seeing the universe. Bell's theorem has taken this even further. It has shown that the universe has a nonlocal aspect and that even if quantum theory is eventually replaced by some deeper theory, this nonlocal feature will always be present in physics.

What nonlocality means remains something of a mystery. There are certain clues from new studies in physics and mathematics that suggest space may be much richer than anyone suspected. There is also the suggestion (made in this book) that nonlocality may extend beyond quantum theory into the entire universe of matter and mind.

Carl Jung has proposed that meaningful patterns in the universe are generated through acausal connections. In an analogous way, the ideas of nonlocality and acausal harmony may extend through the whole of nature to embrace mind, brain, and body. Indeed, as the far-reaching implications of Bell's theorem are unfolded in physics, they may form a new and unified basis for a deeper understanding of the nonlocal universe we live in.

INDEX